Science in History

J. D. Bernal

In 4 volumes

Science in History

Volume 4: The Social Sciences: Conclusion

92 illustrations

The M.I.T. Press
Cambridge, Massachusetts

First published by C. A. Watts & Co. Ltd, 1954
Third edition 1965
Illustrated edition published simultaneously
by C. A. Watts & Co. Ltd and in Pelican Books 1969
Copyright © J. D. Bernal, 1965, 1969

Designed by Gerald Cinamon

Seventh printing, 1986
First MIT Press paperback edition, March 1971

ISBN 0 262 02076 9 (hardcover)
ISBN 0 262 52023 0 (paperback)

Library of Congress catalog card number: 78–136489

Contents

Part 8

Conclusions

Chapter 14: Science and History

Acknowledgements

This book would have been impossible to write without the help of many of my friends and of my colleagues on the staff of Birkbeck College, who have advised me and directed my attention to sources of information.

In particular I would like to thank Dr E. H. S. Burhop, Mr Emile Burns, Professor V. G. Childe, Mr Maurice Cornforth, Mr Cedric Dover, Mr R. Palme Dutt, Dr W. Ehrenberg, Professor B. Farrington, Mr J. L. Fyfe, Mr Christopher Hill, Dr S. Lilley, Mr J. R. Morris, Dr J. Needham, Dr D. R. Newth, Dr M. Ruhemann, Professor G. Thomson and Dona Torr. They have seen and commented on various chapters of the book in its earlier stages, and I have attempted to rewrite them in line with their criticisms. None, however, have seen the final form of the work and they are in no sense responsible for the statements and views I express in it.

I would like also especially to thank my secretary, Miss A. Rimel, and her assistants, Mrs J. Fergusson and Miss R. Clayton, for their help in the technical preparation of the book – a considerable task, as it was almost completely rewritten some six times – and its index.

My thanks are also due to the librarians and their staffs at the Royal Society, The Royal College of Physicians, The University of London, Birkbeck College, The School of Oriental and African Studies, and the Director and Staff of the Science Museum, London.

Finally, I would like to record my gratitude to my assistant, Mr Francis Aprahamian, who has been indefatigable in searching for and collecting the books, quotations and other material for the work and in correcting manuscripts and proofs. Without his help I could never have attempted a book on this scale.

J. D. B.

1954

Acknowledgements to the Illustrated Edition

For the preparation of this special illustrated edition of *Science in History*, I must thank, first of all, Colin Ronan, who chose the illustrations and wrote the captions.

I should also like to thank Anne Murray, who has been responsible for correlating all the modifications involved in producing a four-volume version and for correcting the proofs.

Finally, I thank my personal assistant, Francis Aprahamian, who advised the publishers at all stages of the production of this edition.

J. D. B.
1968

Note

In the first edition of this book, I avoided the use of footnotes. A few notes have been added to subsequent editions and are marked with an asterisk (*) or a dagger (†) (if there is more than one footnote on a page). The notes have been collected together at the end of each volume and are referred to by their page numbers.

The reference numbers in the text relate to the bibliography, which is also to be found at the end of each volume. The bibliography has eight parts that correspond to the eight parts of the book. Volume 1 contains Parts 1–3; Volume 2 contains Parts 4 and 5; Volume 3 contains Part 6; Volume 4 contains Parts 7 and 8.

Part 1 of the bibliography is divided into three sections. The first contains books that cover the whole work, including general histories of science. The second section contains histories of particular sciences and the books relevant to Part 1. The third section lists periodicals to which reference has been made throughout the book.

Parts 2, 3, 4 and 5 of the bibliography are each divided into two sections. The first section in each case contains the more important books relevant to the part, and the second the remainder of the books.

In Part 6 of the bibliography, the first section contains books covering the introduction and Chapter 10, the physical sciences; and the second section, Chapter 11, the biological sciences.

Part 7 of the bibliography contains books covering the introduction and Chapters 12 and 13, the social sciences.

Part 8 of the bibliography contains books covering Chapter 14, the conclusions.

The system of reference is as follows: the first number refers to the part of the bibliography; the second to the number of the book in that part; and the third, when given, to the page in the book referred to. Thus 2.5.56 refers to page 56 of the item numbered 5 in the bibliography for Part 2, i.e. Farrington's *Science in Antiquity*.

The Social Sciences
Past and Present

Introduction to Part 7

Man's knowledge of the society in which he lives has been and still is far more difficult to gain than that of the material world around him or of the plants and animals that live in it. The sciences of society are, as a group, the latest and most imperfect of the sciences, and it is doubtful how far in their present form they can be called sciences at all. Indeed, as has already been pointed out (p. 455), the British tradition of science, embalmed in the Royal Society, refuses to recognize them as such, and the term Science in Britain and America – not elsewhere – is reserved for natural science. Just because the social sciences deal with the springs of human social action, they have a more immediate relation with history than have the natural sciences. It is for that reason that some account of the social sciences must find its place in any comprehensive attempt to relate the development of science in general to that of society.

Because of their late development, the social sciences as a group lacked the autonomy of the natural sciences during the nineteenth century, and carried over into the twentieth many pre-scientific ideas derived from custom and religion. Accordingly to make the social sciences of the present time comprehensible it is necessary to go back over much of the period that has already been covered for other sciences in earlier chapters. For this reason I have thought it better to divide the treatment into two chapters. Of these the first chapter (12) brings the story of the social sciences up to the First World War, and the second chapter (13), deals with more recent times.

Chapter 12 starts with a brief description of the scope and divisions of the social sciences, bringing out the essential differences between them and the physical and biological sciences. The reason for their backwardness is shown to derive not so much from the intrinsic differences or the mere complexity of subject-matter, but from the strong social pressure of established ruling groups to distort, if they cannot prevent, serious discussion of the foundations of society. The way in which this attitude has been broken down successively in social revolutions is next discussed, showing how it has led in our own time to the emergence and development of two rival systems of social science corresponding to the division between *capitalism* and *socialism*. The second system began to take shape in the aftermath of the French Revolution, but only found its full explicit expression in the Communist Manifesto of Karl Marx and Friedrich Engels in 1848.

From 1848 on it will be necessary to discuss the development of these two divergent systems. The first phase of that development, which lay entirely within the framework of a world-wide capitalist economic system, lasted until 1917. That date marks the most appropriate end for this chapter, though it does not correspond precisely to that chosen, for different reasons, for the beginning of the new era in the physical sciences (1895). The later developments of both capitalist and socialist systems of social science are continuous with their earlier forms, but from that date on each finds its fullest expression in a distinct society. These developments and their interaction are the theme of Chapter 13.

I realize that this treatment breaks with the time sequence of the book but it has seemed to me preferable to any alternatives I have considered. Some readers may prefer, as I have indicated in the preface (Volume 1), to take the historic sections of this chapter separately in conjunction with Parts 1 to 5 and to go straight on from Chapter 11 to Chapter 13. Thus 12.2 goes with Part 2, 12.3 with Part 3, 12.4 with Part 4, and 12.5–7 with Part 5. The need to deal with two streams of ideas from 1848 onwards imposes a further difficulty. Here it would have been possible to present together the whole of the sequences of the academic and Marxist social sciences, one after the other. The actual arrangement adopted, dividing each sequence in half at 1917 and discussing them in parallel, seemed more suitable because it kept contemporary ideas more closely together. In any case the presentation of the interaction of historical events and social ideas must take the form of a kind of *counterpoint* in which different themes are worked into a common pattern.

Chapter 12

The Social Sciences in History

12.1 The Scope and Character of the Social Sciences

The social sciences may be divided into two broad groups, the *descriptive* and the *analytic*, though naturally neither term is exclusive. The descriptive social sciences, such as *archaeology, anthropology*, and *sociology*, describe societies, past and present, their structure, interactions, and development. They can all be grouped together into the much wider category of human *history*. The analytic social sciences attempt to discover the underlying relations determining aspects of the behaviour of societies with special emphasis on the societies of today. *Economics*, the science of *law, political science*, and *education* would normally be taken in this category. For reasons that will be set out later they also include much of *psychology* and of *philosophy*, particularly *moral philosophy, ethics*, and *aesthetics*.

SOCIAL AND NATURAL SCIENCES

All these studies can be classed as sciences only in so far as they employ the scientific methods used in the natural sciences, that is, in so far as they rest on a material basis and their accuracy can be checked by successful prediction and practical use. This has been done in the social sciences only to a very limited extent because of intrinsic and special difficulties which will be discussed later. Consequently many parts of them are sciences only by courtesy or for examination purposes: they merge imperceptibly into the non-scientific forms of *religion, literature*, and the *arts*, those human activities concerned with the communication of ideas, images, and feelings, which taken together contribute to the culture of society and assure its vitality and development. Indeed much of the best social science now, as in the past, is found in novels and poems, in plays and films. The relations of social science to practice, that is to the control of society, are also vaguer and more derivative than are those of the natural sciences to the control of the material world. *Business, industrial organization, administration, law*, and *politics* are all practical social activities, but they are still far from being applied social sciences. In fact, much social science is merely the putting of the current practice of the trades and professions into learned language.

It is this association with the activities of groups of interested persons, rather than with an indifferent material world, that has, more than any other factor, prevented the social sciences from acquiring the relatively independent character of the natural sciences. In Marxist terms, whereas the natural sciences are primarily concerned with the *productive forces* of society, the social sciences deal with *productive relations* and the ideological *superstructure* built to maintain and justify them. Even without the Marxist analysis it is apparent that the development of social sciences in the capitalist world has lagged far behind that of the natural sciences. The stage they have reached is indeed somewhat analogous to that reached by the natural sciences before Galileo and Newton. They are essentially discursive and classificatory and, though in modern times they have added measurement in its statistical form, they still lack adequately designed or controlled experiments – the test of practice in application – that established the natural sciences on a firm material basis from the seventeenth century onwards. In common language, the social sciences are fine talk, but they don't work. They are useful for providing subjects for degrees and theses, for teaching posts, and for jobs in advertising and scientific management. Social scientists, however impressive and ornamental, are not yet as indispensable in the capitalist world as chemists or engineers.

REASONS FOR BACKWARDNESS

Now there are many reasons for the backward state of the social sciences and it is of great importance to be clear at the outset as to which are the operative and which are only the apparent and deceptive reasons. In the first place, two reasons of a philosophical nature have been urged to prove the intrinsic impossibility of social sciences at all analogous to the physical sciences. The less important and easiest to refute is the alleged impossibility of experiment in the social sciences. Now it is quite true that without experiment there can be no complete science. What prevents such experiments and observations, however, is not anything inherent in social science but something in the society which it studies. Under capitalism only relatively trivial social experiments can be carried out. The largest such experiment, that of the Tennessee Valley Authority (p. 964), was undertaken only in a time of serious crisis, was hedged with restrictions, and was never imitated elsewhere in the United States, because of its very success.[7.99] The reason hrge-scale experiments cannot normally be carried out is essentially because any such experiment would require the full and free co-operation of the people involved, without the restrictions that are imposed by respect for private property, vested interests, and profits. Such 'experiments' as do take place, for instance in the setting up of social services, are devised and carried out by administrators for politicians, and are so limited in their application to social life that they have little claim to be called scientific. At best they are carried out for the people, never by the people. On the other hand, full-scale social experiment is the rule in the new socialist countries. There, in the development of industry and agriculture, societies deliberately change their habits of life and learn from the results how to plan further change. There also the initiative comes from the workers and farmers themselves, as witness the movements of the Stakhanovites and the worker innovators in the USSR, the Chinese People's Republic and other socialist countries (pp. 1191, 1196 f.).

The second reason given is the claim that the social sciences are intrinsically different because their study involves value judgements foreign to the natural sciences. Such absolute and timeless concepts as justice and beauty are, it is claimed, forever outside the scope of the scientific method. Now this is deliberate obscurantism, made no better by the fact that it derives from the most reputable ancient sources (pp. 195 f.). On the contrary, it is precisely the function of the social sciences to analyse and explain these 'values' in their social and historic context, and to show how they need to change with the social changes of the future (pp. 1307 f.).

Besides these illusory reasons, there remain three others that have a limited validity. The first of these is that the social sciences differ from both the physical and the biological sciences in that man is himself part of the society he is studying, and hence that the observer and the observed become so confused that a really scientific approach is difficult if not impossible.

The second reason given is that as human society is more than the sum of the individuals that make it up, its study must be more complicated than that of human psychology. Consequently, as man is the most complex of animals, the study of man must be far more complicated than that of the biological and physical sciences. In this view, the mere difficulty of the subject is considered sufficient to explain the slow progress in it.

The third reason given is inherent in the changing nature of society. In other sciences progress can take place by a closer and closer approximation to the knowledge of an unchanging or repetitive system of Nature, as in the physical sciences, or a system like organic evolution, where the change is so slow as to be difficult to detect. In society change is rapid, and the social sciences have hardly provided an analysis of a situation before it has turned into a new and different one. The antics of the classical economists in the face of the great slump of 1929 is a sad example of this. They had by the third decade of the twentieth century found what seemed to them an admirable analysis of economic equilibrium brought about by free exchange and unfettered competition. But this situation had already for fifty years had little resemblance to actuality. Imperialism, monopolies, and State restrictions were factors the intrinsic nature of which classical economists could not bear to admit. They tried to treat them as undesirable external obstacles to a free economy, instead of the natural products of its inner evolution. This is no isolated example. The social sciences, which should be in the van of social progress, are in practice in present-day capitalism apparently doomed always to lag years or decades behind the situations they have to analyse.

The cogency of these last three reasons for the special difficulties facing social science cannot be denied. It is, however, highly questionable whether taken together all three can account for the backwardness of social sciences; they have more the air of excuses than reasons. Analogous difficulties have not halted the physical sciences, which have found ways of coping with subjectively distorted, highly complex, and rapidly changing phenomena.

SOCIAL SCIENCE IN THE SERVICE OF ESTABLISHED ORDER

Far more potent than any of these factors is another, intrinsic to society itself, which has operated and still operates most effectively to prevent the formulation of any genuine unbiased social science in the conditions of a class society. The history of the social sciences shows clearly enough that the effective reasons which have held back their development have been strong and positive ones, imposed by those who controlled and principally benefited by the organization of society itself. All through the written history of mankind and, by inference, through a great deal of the unwritten history as well, it has been a very dangerous thing to look too closely into the workings of one's own society. It has always been in the interest of ruling classes to have it generally believed, both by their own members and by their subjects, that the order of society which secured them their privileges has been divinely ordained for all time. With the advent of capitalism and the decline of the age of

faith, such sanctions had lost their full force. For example, the verse
containing the words:

> The rich man in his castle,
> The poor man at his gate,
> God made them high or lowly,
> And ordered their estate

has at length (1950) been tactfully omitted from *Hymns Ancient and
Modern*. Nevertheless, it could still be held that social and legal forms
represented a natural order based on the unchangeable laws of economics. Too much looking into the structure or workings of society might
bring out arbitrary and unjustifiable features which might unsettle
obedient subjects or, in a later time, the free and independent electors.
That is why Plato deliberately constructed myths instead of rational
explanations for the common people in his *Republic* (pp. 193 f.). That
is why in its days of power the Church considered man's duties to his
neighbours and to authority, which then took the place of social science,
to be a branch of theology and its own exclusive concern. That is why,
even now, social sciences are not regarded as a suitable subject to be
taught in the schools (p. 1148).

Characteristically enough, the people who discouraged the pursuit of
the social sciences did not do so because they thought the subject was
philosophically or practically intractable. They did not think the right
answers were difficult to find at all. They knew them already, without
any of the tediousness of searching them out by scientific methods. They
were either self-evident or, when, like the legal 'Acts of God', they were
too absurd for any rational person to think out for himself, they were
divinely revealed.

The image of society presented to common man by his betters
throughout the ages has usually been simple and clear. The pattern of
social behaviour was determined by *custom, institutions,* and *morality*;
or, in others words, by doing what is done or paying the penalty. The
very word 'ethics', on which the subject of good morality is taught in
universities, means no more than 'what is done' or, more properly,
'what people do'. But that pattern could never be, from the very
interests of those propounding it, a *just* pattern, though this fact could
never be admitted. Nor would the theory purporting to support it be a
true or *scientific theory*. In the earlier sections of this book it has been
evident how, at almost every stage, class interests held back and distorted natural sciences. As the social sciences had little or nothing to

contribute directly to material productivity, and since there was a much greater interest in their distortion, it is not surprising that they fared far worse. In short, the backwardness and emptiness of the social sciences are due to the overriding reason that in all class societies they are inevitably *corrupt*. No real science of society can exist that does not begin with the recognition of this fact. Nor can it be fully applied until classes are superseded.

SOCIAL CHANGE ENGENDERING SOCIAL SCIENCE

This is not to say that no social science of any kind is possible in class society, but that it can appear there only by virtue of the changes in that society. At first sight the maintenance of the convention of the blind acceptance of things as they are might seem to be sufficient to prevent social change. But that is reckoning without the evolution of societies themselves, under the influence and the growth of new productive forces and of the consequent new productive relationships. Societies have time and again been torn by struggles between the representatives of old ruling classes, who try to freeze these relationships, and those of new insurgent classes, who must perforce try to break them down as hindrances to their own use of the new productive forces. The central theme of human history has been a sequence of cumulative economic and social strains and rapid releases; it is in the periods of release or revolution that the theories of the nature of society come to be examined and reformulated.

RELIGION AND SOCIAL STRUGGLE

The great world religions arose in disturbed periods in early civilizations, and the questions they strove to answer were fundamentally urgent social questions (pp. 171 f.). Confucius and Lao-tze; Gautama and Mahavira, Zoroaster, the Hebrew Prophets, Jesus Christ, and Mohammed were all active in times of great economic and social transition. They expressed a violent criticism of the society existing in their times, and built up successively new schemes of the rights and duties of mankind. These, it is true, were expressed in religious terms, and the reformers often claimed that they were restoring the just and stable relations of older times. However, in social life there is no turning back, and the great religious reformers became, whether they wished to or not, social innovators. As such they became as surely the founders of social science as the Nature philosophers of Egypt, Babylonia, and Greece were the founders of physical science (pp. 132, 171).

The new social interpretations, the new ideologies, which arose out of the struggles of critical periods, became in their turn powerful weapons in effecting the change to a new order. Knowledge of society is never a passive dogma; it is always active either in preserving or in destroying a social system. The great movements of bourgeois liberation from the restriction of feudalism – the Renaissance, Reformation, the rebellions and revolutions of the sixteenth, seventeenth, and eighteenth centuries – all occurred in periods when the bases of society were put in question. Here, though at the outset the formulation was still in terms of religion, the influence of the new natural sciences began to make itself felt, and social science as a definite scientific discipline can be said to have been born. It was then that the comprehensive though ill-defined doctrines of liberal individualism which were to become the dominant ideology of capitalism first took form.

In the nineteenth century, when the social order of capitalism began to be challenged by the working class in the name of socialism, a new cycle of social criticism and understanding was set in motion. This has gone on with gathering momentum ever since, and is at its height in our own time – one of change and stress unequalled in any era of the past. Never before has the field of the social sciences – the very framework of society and the rights and duties of every individual in it – been the subject of such searching and passionate debate.

Between the years of stress, however, consolidation of privileges takes place and with it a gradual freezing of social science as orthodoxy in religion and politics. This leads to an active discouragement of social inquiry for fear it should become the basis for further social change. In the age in which we live we can observe both the tendencies of conservatism and those of change. In the orbit of so-called Western civilization, that is in capitalist countries on both sides of the Atlantic, we can see the last stages of the general suppression and mystification of the knowledge of society. There is great insistence that the study of society is a pure and objective science, divorced from direct concern with the changes in society. This, though it ostensibly puts the social sciences in the category of the respectable physical and natural sciences, removes from them the possibility of experimental test, the only means of solid advance. Social science becomes an accumulation of harmless platitudes with disconnected empirical additions. Where the social sciences are invoked it is to justify the existing order, either directly by pointing out the essential harmonies of the system, or indirectly by pointing both to the impossibility and the wickedness of any suggestion of changing it.

Meanwhile, not only in the socialist part of the world, in the Soviet

Union, in the People's Democracies, in China, and in Cuba, but also wherever people are coming together to criticize and oppose the oppression of class societies or colonial oppression, new kinds of social science have been growing up. These begin at the other end; they are practical social sciences in which the peoples themselves are *changing* their social relations together with their material environment, and are discovering the principles and mode of operation of society at the same time. This is the first full social *science*, because, as in the case of the other sciences, here also it is only out of practice and through practice that a secure foundation for human knowledge can be laid.

THE HISTORIC ELEMENT
IN SOCIAL AND NATURAL SCIENCES

The fact that the social sciences are and must be closely related with political and economic struggles distinguishes them in degree, though not in kind, from the physical and biological sciences. For, as explained in earlier chapters, the latter are also far more responsive to social factors than has usually been recognized. In another respect also, the fact that they deal with rapidly and irreversibly changing phenomena, the distinction between them and the other sciences, though large, is also only of degree.

Both natural and social sciences change in methods as they advance, but whereas, until very recently, the change in natural science depended only on the increasing march of knowledge, over most of social sciences not only do methods change but the field studied changes even more rapidly. Indeed specifically in history, and in other social sciences to the extent that they are historical, the chief interest is in that very change itself. Though the methods of acquiring facts about the past or present, as developed in the sciences of archaeology or anthropology, may change comparatively slowly, the methods of interpreting those facts change with the experiences gained from the movements in the history of today. This tendency, which has always operated in historical studies (for much of history has been studied largely on account of the support it gives to the protagonists of current controversy), has only been made fully conscious through the influence of Marxism.

The open admission that this is the case is still strongly resisted in most academic circles, while the predilection for the study of ancient history and anthropology has arisen to a very large extent from the desire to escape from the difficulties of having to analyse and study the nature of our own societies effectively, that is practically. We still know far more sociologically and statistically about the economic life of a West African village than we do about that of an industrial town in Britain. That is largely because it had seemed reasonable, at least until recently, to treat the African village as a stable entity to be changed as little as possible, and then only in the interests of the occupying power, which changes, needless to say, must always coincide with the true interest of the natives. On the other hand, the people of the industrial town might want a great deal done if they were made aware of what was possible by an effective social study in which they participated.

This criticism of escapism does not apply fully to the best work in social studies in primitive or vanished communities. Such studies[7.23; 7.175] are already illuminating much of what we take for granted in modern society, showing us to what extent we carry with us 'human nature' or 'self-evident truth' ideas which survive merely because they have been taken over unexamined from the forgotten past, and are found to be most convenient to the dominant class of the present day.[7.42]

SPONTANEOUS CHANGE IN SOCIETY AND NATURE

We are beginning to learn that the *historic* aspect, which is all-important in the social sciences, has still a large part to play in the biological sciences, where we call it evolution, and even in the physical sciences (p. 768). The whole of science has much more in common with the social sciences than we have thought in the past and is, like them, though not so obviously, very much influenced and controlled by the ruling classes of the various societies that have contributed to its development. In this respect the social sciences have another most important contribution to bring to the rest of science. It is in society and only in society that we can easily and repeatedly observe really spontaneous changes at work, that is changes arising from inside a system itself and not merely due to the effect of external influences on it. Over and over again in history, even in the history of our own times, such changes or revolutions have occurred and can be studied. In the past, it is true, for lack of explanation they were often referred to trivial or supernatural origins, but with increasing social awareness it has become more and more apparent that they originated in internal conflicts.

It was the observation of social change in the great disturbances of the mid nineteenth century that led Marx, as we shall see (pp. 1069 f.), to the understanding of the fundamental dialectical nature of spontaneous change in society.

THE PLACE OF THE SOCIAL SCIENCES IN GENERAL SCIENCE

The accepted system of the order of the sciences places the social sciences as the last members of a series that begins with mathematics, runs through physics and chemistry to the biology of animals, to that of man, to psychology, and ultimately to sociology. Scientific knowledge begins, in this view, with the exact sciences and ends with the social sciences. Actually this arrangement disguises and distorts the relation of man to society. The whole scheme has essentially a religious origin, one that treats the creation of man and the ordering of society as divine commands and not as a natural self-development from an animal stage. The first formation of human societies through work in common, with the development of language and traditional techniques, helped to create the psychology of man as we know it and profoundly modified its material basis in brain and body. Certain aspects of psychology, the study of those faculties which man has in common with the animals, such as the acuteness of hearing or the mechanism of image formation in the brain, are essentially part of biology (pp. 944 ff.).

Nevertheless the major part of psychology deals with man as he has been made by society. For the characteristic of mankind as a whole that marks him off from all non-social animal species is the persistence and continuity of social integration. Even the most elaborate of animal societies have nothing approaching the practices of training and tradition that we find in the most primitive human groups. Man cannot properly be said to exist outside society. Each one of us from birth is put through an elaborate system of what is called 'bringing up', with countless traditional performances and responses elicited and fixed by habit at every stage from the cradle to the grave, both, incidentally, social constructs. In a very real sense *man is a self-training animal* (p. 74). Everything we call 'natural' or 'human nature' is wholly a product of human social conditioning. When we say these conditions are permanent, that there is such a thing as an unchanging human nature, we are either repeating the statement, which is untrue, that society itself does not change, or else we are expressing the wish, which is genuine enough on the part of some people, that it should not change. Man's psychology, his wishes, his fears, failings, and virtues, are all part of this self-perpetuating social pattern, continually changing yet remaining an unbroken connexion with that of earlier times. The social sciences are not a group of separate studies, but one single study of one single and growing society, however many and varied its branches may be.

VALUES AS SOCIAL CONSTRUCTS

It is not only psychology that lies in the field of the social sciences. So does the study of the whole body of ancient attitudes that we have called philosophies and religions, with their sub-sections on ethics, morals, and aesthetics. The so-called eternal values sanctified by Plato – the *good*, the *true*, and the *beautiful* – are social constructs and have no meaning apart from society (p. 195). Not only that, they grow and change with society, and all attempts to fix them or elevate them into eternal values are simply attempts to fix the particular forms of society, attempts always doomed to failure.*

This does not mean that no values are real. Every achievement of man in understanding and improving his relations with his fellow men is a positive achievement as real as any technical one. But instead of being, like technique, embodied in material objects, they take form in often even more permanent social institutions. There are good actions, true sayings, and beautiful things, but these are the acts and choices of real men and women, and the values they have come to represent are no empty abstractions but the cumulation of human experience throughout the ages. Though the contributions of the creators of human values – the poets, the builders, the revolutionaries of the past and the present – are not lost, they have become an undying part of a common human tradition.

Nevertheless, because that tradition always grows and changes they can represent no finality to which the future must be bound. What is good, true, or beautiful changes with the times. New and more complex societies make more exacting demands which only the people who live in them can satisfy. At all times in history there is a struggle in society between the forces of the future that will bring new things into being and the forces of the past that strive to stifle them.

The great human creations in morals, in science, and in art are brought into being at each stage of the struggle by the progressive forces. For that very reason they have a common character which still moves and inspires the struggles of later times. In that sense and in that sense only do they represent permanent values. Taken blindly, as established truths, they can in later ages be used to retard social development. Then they have to be fought against and broken; new wine must not be poured into old bottles. It is only by accepting the fact of the change of attitude towards values in society that we can understand the deeper and more permanent nature of the values themselves, and use them fruitfully and harmoniously in building a new society.

THE METHODS OF THE SOCIAL SCIENCES

It is not only in subject matter but also in method that the social sciences differ from the natural sciences. In this respect social sciences have been most badly served. Three different and incompatible methods – literary, biological, and mathematical – have been successively, and are now being simultaneously, applied to the study of social problems. As will be shown in later sections, the social sciences have suffered from not being treated by methods suited to the complex and self-involved nature of human societies. By attempting directly to apply the methods of other sciences, especially biology, oversimplified, false, and dangerous conclusions have been reached. The use of statistics has often given a delusive appearance of accuracy to social data.

The disputes, inconsistencies, and vagueness of the social scientists have made the general public uncomfortably aware that social science is not quite the same thing as natural science. The attempt to pretend that it is so is premature, at best an illusion, but sometimes a conscious deception. The effect of teaching social science without realizing this renders such teaching a waste of time; it creates a false sense of achieved knowledge, and in doing so actually prevents the student from seeing the most elementary things that he might have hit on himself without any science at all.

The existence of classes and the exploitation of the poor by the rich have been for 4,000 years the most outstanding fact of social life. Yet in the 'science' of society far greater efforts have been made to pass it over or explain it away than to study it and work out the consequences of the fact itself. What social science needs is less use of elaborate techniques and more courage to tackle, rather than dodge, the central issues. But to demand that is to ignore the social reasons that have made social science what it is. To understand that we need first to look more deeply into its history.

12.2 The History of the Social Sciences

SOCIAL SCIENCE IN ANTIQUITY

The preceding paragraphs contain a brief and formal analysis of the social sciences and of their relation to the natural sciences and to other aspects of culture. To be complete such an analysis requires, especially in the case of the social sciences, a study of their history in relation to general historical developments. Now this relationship, which it is the major purpose of the book to discuss, is a very different one for the natural and for the social sciences. On the one hand the social sciences have contributed little or nothing directly to the changes in productive methods on which all permanent improvements in man's lot depend; on the other they are linked far more closely than are the natural sciences to changes in the economic and political institutions of society. They are also more evidently related to pre-scientific traditional ideologies of religion and philosophy. For that reason to understand them fully it is necessary to go back to the very beginning of human society.

THE SOCIAL SCIENCE OF EARLY MAN: RITUAL AND MYTH

The origin of man and of society has been discussed in an earlier chapter (pp. 65 f.). Man's own concepts of that origin, arising from ritual and myth, could be expressed only in terms of the society he knew. There are innumerable examples of this in the myths of existing savages (pp. 85 f.), and we may here fairly presume the same for primitive man, as his material remains indicate that he practised very similar rituals. These recalled, through imitation or symbolism, the important events in actual life and implied the practical belief in the magical control of Nature by man. The purpose of the ritual and its accompanying songs and myths was in the first place to secure food or other desirable goods. Almost as important, and not clearly distinguishable from it, was the maintenance of the social pattern. The common *rites de passage* from one stage of life to another: the birth rites, the initiation rites, the marriage rites, and the funerals which are still enshrined in the *sacraments*, all were dramatic representations and fixings of the necessary relations of social life. The importance and the strength of myth and ritual are shown by the fact that, in spite of a transformation to quite a different way of life, a great deal of this old myth and ritual is still with us, embalmed in the structure of the surviving religions of today and also in the common customs that have outlasted the religions of the past.

FROM MYTH TO MORALITY

But as society became more complex and classes arose such myths were found to be inadequate. It is fascinating to trace the transition, with civilization, of the early magical and mythical analysis of society into one which becomes moral and rational, one in other words corresponding more to the organizations and technical necessities of city life (p. 108). The process is a natural one, as is shown by its parallel development in ancient China, in India, in the Near East, and even in the civilizations of the Americas.[7.168] In all cases the old pictures are not destroyed, but they are given a new meaning. The original codes of behaviour of the Hebrews, now to be found in the book of Leviticus, were largely primitive tribal taboos and magical precepts.[7.136] The ten commandments that we know from the book of Exodus are the result of later changes, when the moralizing influence of civilization had had time to prey on the older irrational tribal customs.

The order of the transformation is that in the first place an action such as eating a totem animal is avoided because it is not done (*ne-fas*, fatal), or more precisely because it is firmly believed that it will automatically bring disease or starvation to the tribe. In barbarian times this operation of natural forces becomes personalized in the form that 'such and such actions are unlucky and will remove the favour of the gods' from the individuals practising them. It is only when we come to civilization that we find that such and such an action is considered inherently immoral. In the process the more picturesque elements tend either to be dropped out altogether or retained only in an allegorical sense, ultimately to become mere ornaments, like the paper horses burnt at Chinese funerals or our own superstition about spilling salt.

We must not delude ourselves into thinking that this corresponded with any actual improvement in morals; in fact rather the contrary was the case. Savage culture is sufficiently equalitarian not to require either the internal sanctions of morals or the external sanctions of law. Laws as such appear only when class distinctions have hardened and property and privilege must be protected. The discovery of a law prohibiting a certain offence is not a proof that such offences were not committed, but rather that they were on such a large scale that they could no longer be ignored. The rise of a conscious morality corresponds to the transition, already discussed, from a tribal classless society, with goods in common, to a class society in which property is privately appropriated.

The growth of class societies also corresponded, as we have already seen (p. 96), with the transformation into primitive science of that part of ritual and magic that dealt with the control of the external material world. A theory of society, embodied in religious myths, was necessary only when it was needed to explain and defend an inequitable system. Official morality, with its parallel in laws, is intrinsically a two-faced device. On one side it exhorts the majority of the population, that is the poor and humble, to refrain from actions which would incommode the ruling classes; on the other it creates an aura of respect for those ruling classes by setting up an ideal of right and justice calculated to satisfy the people that oppression will be kept within bounds. It is interesting to note how many of the precepts of the Bible are concerned with exhorting the rich to limit their rapacity.[7.136] As class societies became decadent and conflict-ridden, official morality tended to rest increasingly on religion, in which imaginary gods, symbolized by their statues or emblems, were regarded with more awe than the wealthy chiefs or priests in whose image they had originally been made.

THE PHILOSOPHY OF THE GENTLEMAN

It is at this stage too, in the social just as much as in the physical sciences, that the split occurred between the theoretical and the practical, between the book-learning of the ruler and the traditional lore of the ruled, and this is most apparent in the philosophies of the early civilizations of India, China, and Greece. The split appears between the philosophy and knowledge needed by the ruler – the superior man or gentleman of Confucius, the Brahmin of India, or the philosopher of Greece – and the simpler knowledge and more prescribed practice of the common people (pp. 171 f.). In these three centres of civilization (and probably also in Babylonia and Egypt, though we know less about them) conscious attempts were made to formulate a systematic body of knowledge of society for the benefit of those intending or being called upon to rule the state. Such an account is to be found in the Chinese classics, particularly the classic referred to as the 'Great Learning', which states explicitly that the main value of philosophy is as a guide to good ruling:

Things have their roots and branches, human affairs their endings as well as beginnings. So to know what comes first and what comes afterwards leads one near to the Way. The men of old who wished to shine with the illustrious power of personality throughout the Great Society, first had to govern their own States efficiently. Wishing to do this, they first had to make an ordered harmony in their own families. Wishing to do this, they first had to cultivate their individual selves (*hsiu shen*). Wishing to do this, they first had to put their minds right. Wishing to do this, they first had to make their purposes genuine. Wishing to do this, they first had to extend their knowledge to the utmost. Such extension of knowledge consists in *appreciating the nature of things*. For with the appreciation of the nature of things knowledge reaches its height. With the completion of knowledge purposes become genuine. With purposes genuine the mind becomes right. With the mind right the individual self comes into flower (*shen hsiu*). With the self in flower the family becomes an ordered harmony. With the families ordered harmonies the State is efficiently governed. With states efficiently governed the Great Society is at peace.[7.77]

Essentially similar in character was the Plato of the *Republic* and the *Laws*. Both were essentially rational and practical guides to a cultivated aristocracy (p. 194).

The moral and intellectual ideal held up by the Chinese and Greek philosophers is that of the *superior* man or *gentleman*, not himself a ruler but an adviser to rulers, who knows what should be done and is willing, for a consideration of being respected and well looked after materially, to inform the prince of the admirable behaviour of his ancestors or of the line of action indicated by pure reason. This concept

307. The yogi escapes from society and from the world by meditation and subjugation of bodily desires. The teaching contains a kind of negative social knowledge. This photograph of an old yogi was taken at Benares (Varanasi).

of a practical social science, including especially history, philosophy, and a knowledge of the classics as the basis of an education of upper-class youths in the art of ruling, has passed on almost unchanged from the Greeks to the present day. It was fitted to take the place in a new civilization of the old classical education in the youth of Persian nobility – to ride, shoot the bow, and tell the truth.

MYSTICISM AND THE REJECTION OF SOCIETY

The Indian solution, which was adapted to a more rigidly stratified society, was somewhat different, as were those of the Chinese *Taoists*, the *Zoroastrians*, and the mystery religions of the West. Here knowledge was sought as a way of escaping from society rather than attempting to control it. The learning of the ascetics or *yogis* was not thought of as social knowledge, though it was in fact a kind of negative social knowledge, an analysis of society in order to show how hopeless it was to do anything about it, one leading essentially to withdrawal, inaction, and a longing for non-existence (pp. 179 f.). In effect the ascetics and mystics were not socially neutral; by their very existence they excused, almost invited, the maintenance of oppression and misery by unenlightened rulers whose actions might, like drought or pestilence, be considered as part of an inscrutable divine plan. The idea which the ascetics strove consciously or unconsciously to impress on people and rulers alike was that holy men acted as spiritual lightning conductors, protecting an evil society from the wrath of God.

THE BIBLE AND THE PEOPLE

A different, though still religious, solution was the attempt to defend, as by divine sanction, some at least of the characters of a clan society against the debt and slavery which were the inevitable concomitants of class differentiation. Such attempts must often have been made, but only that of the Jews was to have a lasting effect in world history. This was due in part to its incorporation in written form in the Bible, in part to the firm retention of customs by Jews scattered throughout the world, and ultimately to the incorporation of much of the Jewish tradition in both Christianity and Islam. The position of the Jews in the ancient world has already been discussed (pp. 157 f.). They had to maintain a continuous struggle for existence against more powerful, more civilized, and more economically developed neighbours. In this struggle the princes and wealthy Jews were always tempted to collaborate with the foreigners. Against them time and again the people, inspired by the denunciations of the prophets and fortified by the Law, put up a dogged and effective resistance.[7.136] It is especially in the book of Deuteronomy, with its insistence on the liberation of slaves and limitation of debt, that we find a social code which at least limits and mitigates the total brutality of class rule, as exemplified for instance in Babylonian or Roman law.

The resistance of the Jewish people was an isolated one; it depended too much on their belief in their peculiar covenant with Jehovah for it to be able to spread directly to other people. But the same conditions gave rise to a similar, if less explicit and effective resistance of exploited peoples everywhere. The history of antiquity, despite its upper-class character, is forced to record a sequence of agrarian struggles, democratic insurrections, and slave revolts where the impetus certainly came from below. As civilization spread there grew up, in opposition to the superior science of the philosophers, the practical social knowledge of the great mass of the population, the poor and oppressed. It was with the people rather than with the princes or the priests that the true repository of morality was to be found. This was a positive morality of forbearance, comradeship, and mutual help, uncorrupted by the anxieties of maintaining the sanctity of the class system. The philosophy of the poor was not a conscious literary or philosophic creation, but it was embodied nevertheless in well-maintained customs, and was recalled and kept alive in the thousands of songs and proverbs of the people, not all of them expressing the highest respect for their betters.

THE GREEK ANALYSIS OF SOCIETY

Neither the religious nor the popular expressions of social consciousness may properly be called science. They lacked coherence, terminology, and logic. As in all other fields of thought, it was the Greeks who produced the first analytical and logical presentation of the social sciences. Indeed we owe to them the whole terminology of the subject: ethics, economics, politics, and history itself are all Greek terms. The debates, the revolutions, the wars of the Greek city states all turned on social questions, in which the division of the classes was the most prominent. To the Greeks, man was primarily a citizen – the *zoon politikon*, the political animal of Aristotle – though the more primitive classless tribesman was just below the surface. The problem of city government dominated social science, indeed it led to the first systematic observations in the field, such as the collection and comparison by Aristotle and his research workers of the constitutions of 158 cities (p. 207).

His objective, intrinsically unattainable, was to find some system which should provide social harmony while not abandoning class privileges. This was indeed the purpose of his doctrine of the *mean*. Moreover, the Greeks were not limited to comparing one of their own cities with another; they also travelled and traded with peoples of many different grades of social organization, from primitive tribes to the highly organized empires of the East (Herodotus is the father of anthropology as well as history). Nevertheless, like the British in their time, the Greeks had a feeling of the natural superiority of their own city states; the others were just barbarians, people who said ba-ba and could not even talk properly, and whose customs were to be avoided rather than copied. This knowledge of a socially different outside world did, however, give Greek thinkers an objectivity which the Indians and Chinese lacked.

The major contribution of the Greeks to social science was their success in abstraction, in finding words to express common elements in diverse situations without always having to refer to particular instances. This made discussion possible, but it also made it far too easy to use abstract words as if they referred to self-subsistent things, and by an abuse of logic to draw from them conclusions to fit any preconceptions. This was a most deliberate practice, as the Socratic dialogues show, and was aimed at putting discussions of actual cases out of court. If, as in the *Republic*,[7.130] the ideal of justice is abstracted from the conditions of the city where it is dispensed, it is possible to justify in its name the most arbitrary and undemocratic constitutions. This abuse of abstraction was somewhat mitigated in the physical sciences, where at least abstractions could be counted and measured. In the social sciences abstract categories have proved to be unmitigated nuisances and obstructions. The values and ideals that the Greeks had words for still plague us to this day.

Nevertheless, as in the case of the physical sciences, any serious attempt to formulate a scientific study of society must henceforth stem from Greek sources. Before, however, they can safely be used it is necessary to understand how much of the social philosophy of the Greeks was an attempt to justify privilege and slavery, and this bias needs to be heavily discounted in attempting to apply it to current problems. The Greeks can still be most dangerous masters. Marx, it is true, began his intellectual work with his thesis on the atomic philosophy of Democritus and Epicurus. But these were disturbing and radical teachers, outside the main stream of Greek thought, though fated, as we have seen, to inspire the whole of modern science. In the history of the social sciences a major task has been to break the chains of conformist belief enshrined in the work of Plato and Aristotle, a task which has taken much longer in the social than the natural sciences and which is not yet complete.

ROMAN LAW

The Romans, for all their unimaginativeness in natural science (pp. 226 f.), had much more practical knowledge of social science. It was knowledge acquired the hard way, first in stifling by force or settling by compromise the class conflicts in their own city throughout the long struggle between patricians and plebeians, and then in conquering, exploiting, and administering the Empire. That knowledge was expressed not so much in philosophy as in *law*. Roman law was the most complete codification of the conditions of domination of society by the holders of money and political power. In Roman law the rights of property were paramount. Property included, and indeed largely consisted of, slaves, so that the most appalling private injustice could be approved in the name of public justice. Nevertheless it did represent an ordered setting out of social relations (p. 230). Roman law did not survive, as an operating code, the administrative system in which it was

308. Section of a mosaic at Ravenna depicting Justinian (centre).

formed, and became a dead letter in barbarian times. Nevertheless it remained for centuries the framework in which the literate could see society in a rational way. The study of law was to be one of the ways in which the nature of society was to be rediscovered after the Dark Ages.

12.3 Social Science in the Age of Feudalism

SOCIAL SCIENCE AND THE EARLY CHURCHES

It was, however, a different but allied system of social thought that was to take the place of law and order after the disappearance of the Empire. As we have seen (p. 254), the break-up of the classical, mainly Mediterranean, civilizations coincided with the rise of organized religion, an essentially new feature in human history. This was in the first place a more basically democratic system based on the congregation of the faithful. The model was the *Synagogue* of the Jews, originally a rallying place of the resistance movement, clinging to its law and ritual as a protection against the power of foreign rulers (*n.* p. 255). The early Christian *Churches*, though they grew up as revolutionary and communist in the tradition of their founder, had to build up their social philosophy from the elements they found around them, which were largely Jewish or Greek. The need to defend Christianity against both the ideas and the physical persecution of the heathen world forced the separate congregations to unify their beliefs as they strengthened their organization, though heresy and schism prevented them from presenting at any time a completely united front.[7.119; 7.137] When, from economic collapse and barbarian invasion, imperial authority and secular culture were at their lowest ebb in the Western Empire, the Church was able to take over, and to control and regulate almost every aspect of social life in an organized way.

Where this collapse had not occurred, as in the Eastern Empire, neither the Christian Church nor its successor in Asia and Africa, the faith of Islam, ever acquired any corresponding central organization, nor were they able to exert the same political force. Nevertheless, even in these countries, and in India and China as well, religions tended to organize themselves, though on a looser and more local basis. In Islamic countries a great coherence of orthodox doctrine and law was achieved through the institution of universities (*madrasah*), originally schools attached to a mosque. Buddhism developed its own monastic organization in the lamaseries by means of which it spread through most of eastern Asia. It was, however, only in backward areas such as Tibet and Mongolia that it led to the establishment of a religious hierarchy as powerful as that of the Pope. Brahminism remained the least organized of all, with no fixity of belief or cult; but through the institution of caste, of village and family priests and of pilgrimages, it maintained an extraordinarily persistent pattern of culture in its native land, though for the same reasons it failed to spread abroad.

CREED AND FAITH

The establishment of organized religion put an end to fluidity of belief; it also ensured a set of institutional bases for definite theories of society. Indeed religious orthodoxy contained a complete scheme of social science, which, though taken over from mere secular practice, was before long firmly believed to have been laid down by divine order once and for all. Modern social science does not therefore begin from an unprejudiced examination of society, but from modifications, at first tentative and conforming and becoming gradually more bold and heretical, of the religious picture of society.

THE CITY OF GOD

Even religious orthodoxy, for all its apparent permanence, has its history. The Church had in fact largely to remake social science in the West. The heritage of the classical philosophers and jurists needed considerable adaptation to the conditions of the decaying Roman Empire, or of the barbarian kingdoms that followed it. One aspect of the change was the need to bring into harmony the concept of the individual detached soul escaping by an ascetic withdrawal from the wicked world, an idea derived from the old mystery cults and powerfully reinforced by Manichaeism (p. 268), and that of the recovery and maintenance of order in this world, so that at least the Church could flourish in security. This necessitated investigations both of the nature of the soul – *psychology*, though the word is a later one – and of the framework of society, the divine ordering of the temporal world. It is interesting that the first of the classical authors who is known to us to have occupied himself with the problem of the upbringing of small children was St Augustine, and his observations and recollections of childhood went far to confirm him in his theological support of the doctrine of original sin. It was also St Augustine who adapted the Greek Platonic ideal to the conditions of a decaying classical civilization. His *City of God*, a spiritual consolation for the fall of Rome in

309. The City of God as envisaged by St Augustine. Drawn from a translation made by Raoul de Presles in the fifteenth century. At the top are the saints already received into heaven, and in the seven compartments below are shown those who prepare for heaven by exercising Christian virtues, or who are for ever excluded by the seven deadly sins.

A.D. 410, was the first conscious plan on which the medieval world was to be built (p. 260).

The social sciences of the Muslims suffered in the attempt to reconcile the Koran with the complexities of the life of the conquered cities (pp. 269, 299). Most Islamic philosophers treated social subjects with an infusion of Platonic and Aristotelian ideas (p. 1044).

THE SOCIOLOGY OF THE SCHOOLMEN

The great medieval scholastic disputations were largely about social questions, particularly questions of government, as for example on the respective spheres of Pope and Emperor. St Thomas Aquinas' *Summa Theologica* lays down the character of a just society, conformable to scripture and reason (p. 300), while Dante in his *Divina Commedia* and *De Monarchia* presents this world view as a background to the turbulent life of Italian cities.

The general conception of both saint and poet was in harmony with an integral and hierarchical society, so well established as to seem natural, in which the parts were mutually dependent and in which everyone had his right place. It was in essence a Christianized Aristotelianism. The terrestrial social universe corresponded point by point with the great celestial universe (pp. 306 ff.). Angels of appropriate rank took their station beside individual men and ruled the stellar spheres. This system was explicitly also a changeless and durable one, subject to God's will. It was founded at the creation and it would endure till the last trumpeter had sounded, when the earth would disappear and the celestial hierarchy would be fixed for ever.

HERESY AND SOCIAL CRITICISM

This ideal did not long, if ever, correspond to reality. The actual Church tended to accumulate wealth and become itself the largest of feudal institutions. Criticism of religious belief came the more easily from being linked with that of the un-Christian behaviour of the clergy, and of their exploitation of peasants and the lower bourgeoisie of the towns. The heretics tended in one way or another to go back to early communist Christianity, with a strong infusion of the prophetic books of the Bible. At first the heretics were limited to the more economically advanced regions around the Mediterranean, and were easily put down by crusades and inquisitions. In the latter Middle Ages, with the peasant revolts in England and France and with the initially successful rising of the Hussites in Czechoslovakia, their suppression proved a more

difficult and ultimately a hopeless task. Despite their religious formulation, all these heretical views were originally based on a criticism of the
injustice of the class system of feudal Europe, but they were unable by
themselves to break it down, because they had nothing essentially
different to put in its place; that required the operation of more powerful economic forces.

12.4 Social Science and the Birth of Capitalism

THE RENAISSANCE TRANSFORMATION

These forces were to be found in the growing strength of the towns
and of the new commerce and manufacture, now in the hands of the
rising bourgeoisie, with quite different ideas (p. 379). Just as the system
of the heavens could not satisfy the navigational needs of the new commercial age of the Renaissance, so the social philosophy of the Middle

310. 'The Money Changers' by David Teniers (1610–90). Now in the National
Gallery, this depicts one aspect of the handling of money for profit and interest.

Ages could not satisfy its economic demands. An economy based on land and the traditional exaction of personal service had to give way to one based on trade and small-scale manufacture. The use of money, or rather the charge for using it, had to change from the *sin of usury* to the meritorious *lending at interest*. Services were no longer a matter of loyalty secured by title to land, but had to be bought and paid for. A man was worth what he could make.

While clinging to the forms of religion and even claiming to go back to an older and purer one, the reformers destroyed the whole concept of a unitary social system. What they had to put in its place was an atomic one, a concept of society as an assembly of individuals each of whom had to make his separate treaty with God, and achieve salvation, either by faith or by predestination, according to whether he followed Luther or Calvin. In this respect the Catholic Church came ultimately to agree with the Reformed. Although it held out on theological dogmas, the Council of Trent (1545-63) went a long way to accepting the individual view of salvation through the doctrine of grace (p. 384), and with it went the abandonment of the integral social theory of the medieval Church.

The medieval world-picture was built into so much of the structure and custom of society that it took the efforts of many thinkers a long time to build a consistent picture of the new individualist world. Machiavelli was the most notable of the men of the Renaissance to look at society as clearly and dispassionately as the painters had looked at Nature. Patriotic and essentially democratic Florentine though he was, he could see nothing in his time that could succeed except a carefully balanced play of self-interest, force, and cunning. But even then it did not pay to put down in writing the principles on which so many of the great and pious of history were acting, and all it earned him was a bad name which has lasted to this day. A hundred years before him, one of the last of the great Muslim thinkers, Ibn Khaldun (p. 273), prefaced his treatise on history with a theory of economically determined social development, thus foreshadowing Vico and Marx.

REFORMATION AND REVOLT

The early reformers were by no means a socially homogeneous group. The leaders, like Luther and Calvin, came to be associated with the princes or with the wealthy bourgeoisie of the big towns. The mass support of reform came from much smaller folk, artisans and peasants, who wished the reform to be as evident in this world as in the next. They tended to mix their criticism of the wickedness of the rule of the Pope

311. 'A priest called John Ball stirs up great commotion in England'. From the
Chronicles of Jean Froissart (1337–*c*. 1410). Ball, who took a prominent part in
Wat Tyler's rebellion, was executed at St Albans in 1381. He preached equality of
gentry and villeins, preaching at Blackheath on the text

> 'When Adam dalf, and Eve span,
> Who was thanne a gentilman?'

and his servants with one of the whole system of oppression of the poor
by the rich. This led to peasant and artisan revolts in Germany and
Hungary which, after some successes such as the setting up of an
Anabaptist commune in Münster, were repressed with even greater
ferocity than those of earlier times.

It was from that age, the early sixteenth century, that the idea of a
deliberately instituted Communist State, derived in part from Plato and
in part from early Christianity, began to take shape. The most famous
of these accounts, and the one which has given all of them their name,
was the *Utopia* of the humanist Sir Thomas More. Though he was of
the new upper bourgeoisie, he was disgusted by their greed, and tried
to find a way of avoiding through a communist society the corruption
of wealth and power. It is in keeping with his character that he died in
defence of the old faith.[7.22] Almost contemporary, though in a very

different vein, was the work of Rabelais, a doctor of medicine and a boisterous critic of all pedantry and intellectual relics of the Middle Ages. His *Gargantua* was at the same time a programme of the new humanistic freedom (p. 385). *Fais ce que voudra* is a distant anticipation of *laisser-faire*. Through all his joviality there ran a serious vein. His phrase *Science sans conscience n'est que ruine de l'âme*, though intended for the venal humanists of his time, has its message today.[7.133]

The later humanists, chastened by the miseries and disillusions resulting from the wars of religion, had lost the gay optimism of the high Renaissance. But the very experiences they had been through gave them a more balanced view of the nature of society. Cervantes (1547–1616), a broken Spanish soldier, writes the epitaph of feudalism in *Don Quixote*. The same questioning runs through the plays, particularly the later plays, of Shakespeare. Montaigne (1533–92), a Gascon country gentleman whose highest State office was to be Mayor of Bordeaux, expresses the most reflective thought of the transition to the bourgeois era. In their work is to be found more social science than in that of all the moralists or philosophers of the time. The majority of Renaissance philosophers, dependent as they were on princely favour, deliberately and prudently avoided social questions and concentrated on the individual man's attack on Nature, though with Bacon and his followers there was also the recognition that 'just as merchant adventurers fared better in companies so might philosophers' pp. 41 f.).

NATURAL LAW

In the transition from the religious view of society to the commercial one a large contribution was made by lawyers. Their very profession forced them to undertake much of the work of adapting the whole legal and political system to the needs of the new economy. To them, however, it seemed that they were merely restoring the eternal principles of *natural law*, free from the accretions of barbarous ages. The sixteenth- and seventeenth-century jurists were the heirs of the humanists, and the most distinguished of them always strove to soften the violences of religious antagonism in the service of a cultured and tolerant society, preserving, however, the sacred rights of property. Bodin (*c.* 1530–96), a Frenchman, laid the foundation of a science of history and understood how closely it linked with economics. He was the first to explain the nature of the great inflation of the sixteenth century.

The greatest of the lawyer philosophers was Grotius (1583–1645), the founder of international law. This had become important by the beginning of the seventeenth century, not only because of the establishment of sovereign States, but even more because of the extension of trade over the whole world, touching countries with quite different civilizations. Grotius' own contribution, indeed, started with a brief he held for the Dutch East India Company.* Later, after escaping from life imprisonment to which he had been condemned for supporting religious toleration, he was led to reformulate the whole basis of a natural law independent of Church and State.

THE BIRTH OF BOURGEOIS SOCIAL SCIENCE

The real new birth of bourgeois social science came with the great religious, national, and class struggles of the late sixteenth and middle seventeenth centuries.[7.179; 7.69] To justify rebellion against the King of Spain,† or the cutting off of King Charles' head, necessitated an inquiry into the ultimate purposes of society. The social criticisms of Lilburne and Winstanley[7.181] and the actions of their fellow Levellers and Diggers were the earlier forms of a combined theoretical study and practical reform of the social order. But that movement was premature, and subsided in the general European reaction of the latter part of the seventeenth century. More acceptable for the time was the analysis by Hobbes (1588-1679) of the collective nature of society, leading to his demand for a strong government to control the corporate Leviathan.[7.75]

POLITICAL GEOMETRY AND ARITHMETIC

The great outburst of scientific activity in the middle of the seventeenth century, culminating in the triumph of the new experimental philosophy, could hardly be confined to the natural sciences, though these provided the central inspiration and coloured the approach to other fields of knowledge. It seemed that the methods of measurement and geometrical demonstration that had succeeded so well in the field of physics might do the same in that of society. The immediate results were, however, disappointing. Even the greatest philosophers, such as Spinoza and Leibniz‡ (p. 518), did not succeed in convincing many that propositions about ethics and morals could be proved as rigorously as Euclid did those of geometry.

On the inductive side, however, a start was made in applying measurement to social factors which was to prove of great importance in the future. Graunt (1620-74), a tradesman of London, published his *Remarks on the Bills of Mortality*; for this he was made a Fellow of the Royal Society with the express approval of the King. This was the beginning of vital statistics. He was followed by others such as Halley, who drew up life-tables which the great administrator Cornelius de Witt

312. Bills of mortality were prepared regularly from 1563 in England, but were not necessarily published. At the time of this Bill (1665) about 130 parishes were included, and the Bill was issued every Thursday: cost, 4 shillings a year.

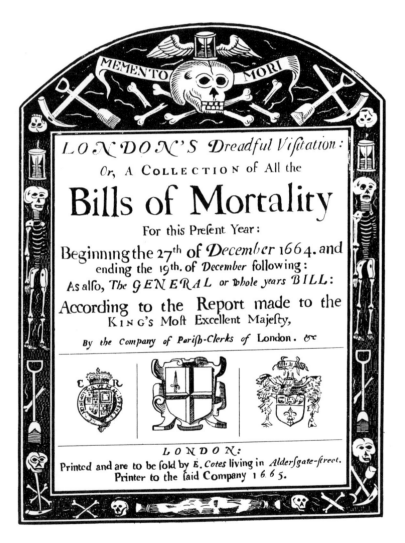

(1623–72) used for selling annuities profitably, thus rescuing the finances of the Dutch Republic. From this springs the whole great business of insurance. Another of the Fellows, the extremely successful lawyer, land-grabber and business man, once secretary to Hobbes, Sir William Petty, started another kind of social science, that of economic statistics, which is so popular today, with his *Political Arithmetic*.[7.160]

12.5 Enlightenment and Revolution

NEWTON AND LOCKE

Political and economic theory became the major social sciences of the eighteenth century. With their development the relations between the physical and the social sciences again became very close. John Locke (1633–1704), the personal friend of Newton, himself a scientist and practising doctor, turned to the new ideas of science to justify the very type of compromise government that the glorious revolution of 1688 had set up. This government effectively gave all power to the city merchants and the landed aristocrats, who together formed the new wealthy bourgeois class. All they wanted was security from arbitrary royal interference or from usurpation from below, and they were prepared to govern in a civil and legal manner so long as they could make and administer the laws themselves. Locke himself helped to found the Council of Trade in 1696 (later the Board of Trade), the first organized attempt to apply the new mathematical methods to public business. It was extremely convenient to have discovered that the universe itself also ran on eternal laws and that with a good constitution there was no reason why anything should change again.

THE WEALTH OF NATIONS

The excellence, in a negative way, of the British constitution was also brought out by the study of economics. Many eighteenth-century writers concerned themselves with economic matters. Hume (1711–76), the founder of the sceptical or agnostic school in philosophy, was definite enough in support of the new institutions of capitalism. He recognized the importance of competition between merchants in lowering profits and keeping down the rate of interest, which favoured even greater commerce. He understood too how increasing stocks of money depressed real wages while keeping profits high.[7.139] Mandeville (*c.* 1670–1733), with his *Fable of the Bees*, suggested that the welfare of the community might be a consequence of the market created by the vice and extravagance of the rich.

The beginning of liberal economics as a serious study was to come only with Adam Smith, one of the band of Scottish intellectuals whose activity coincided with the transformation of Scotland from a miserably backward agricultural country into a centre of industry (p. 530). Adam Smith was impressed with the great prosperity of Britain, which he saw arose from the spontaneous and unorganized development of manufacture which immediately preceded the Industrial Revolution (pp. 509 f.). He saw it as the product of the *division of labour* in industry and of the ease of *exchange* of commodities and manufactures. At the same time he could not fail to note the interference that government restrictions imposed both on manufacture and trade. He saw particularly in the prevailing system of *mercantilism*, which monopolized colonial trade in the interests of a few wealthy merchants, the most serious enemy to the free development of the new forces.[7.38] In support of these views Adam Smith set himself the task of analysing the whole productive-distributive mechanism of society, and embodied the results of his analysis in his *Inquiry into the Nature and Causes of the Wealth of Nations* (1776). This book, which was from its first publication to become the bible of the new industrial capitalism, is one of the great synthetic social testaments, comparable with Aquinas' *Summa* and superseded only by

313. In the eighteenth century, increasing specialization occurred in industry. A tannery from *The Complete Dictionary of Arts and Sciences* edited by T. H. Croker, T. Williams and S. Clark, London, 1764-6.

314. Adam Smith's *Inquiry into the Nature and Causes of the Wealth of Nations,* published in 1776, was an economic treatise of the greatest importance, conceiving as it did *laisser-faire* economics as part of the natural order. The title page is shown here.

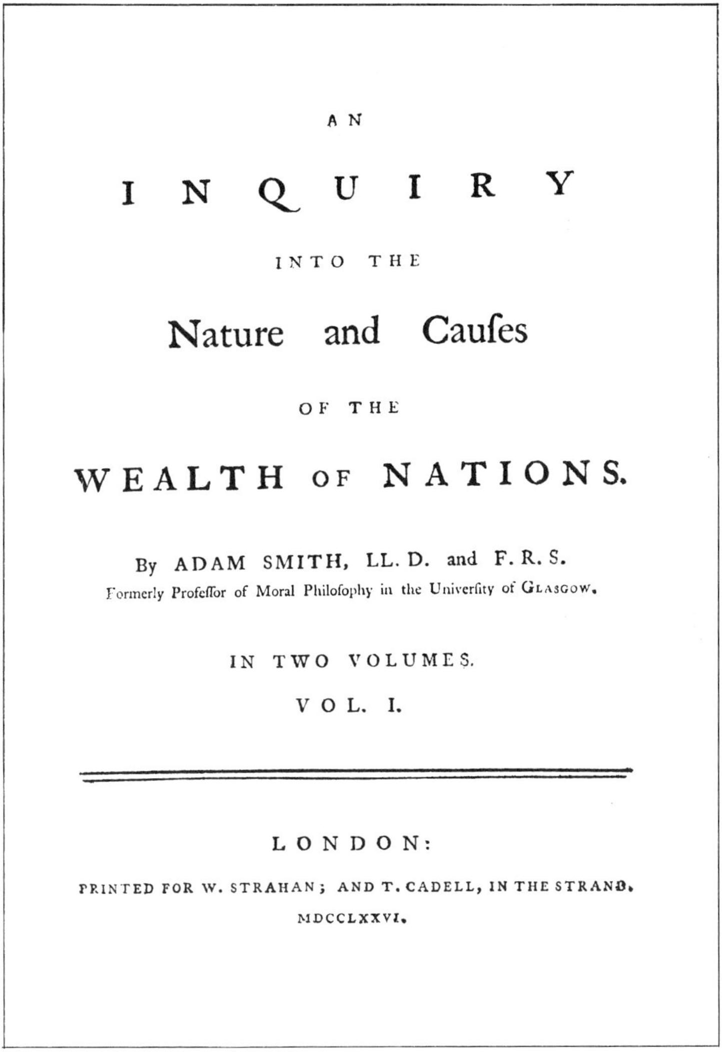

A N

I N Q U I R Y

INTO THE

Nature and Caufes

OF THE

WEALTH OF NATIONS.

By ADAM SMITH, LL. D. and F. R. S.

Formerly Profeffor of Moral Philofophy in the Univerfity of GLASGOW.

IN TWO VOLUMES.

VOL. I.

LONDON:

PRINTED FOR W. STRAHAN; AND T. CADELL, IN THE STRAND.

MDCCLXXVI.

Marx's *Capital*. It is, however, far narrower in scope and intent than either of these. It deals primarily with a new kind of being, *economic man*, a creature living by labour and the exchange of his products with his fellow men, and always making the most advantageous terms he can. Adam Smith explained how these activities had always been restricted in the past by ancient custom, feudal rights, or mercantilist regulators. Now at last he saw in this new enlightened age the prospect of achieving a *natural order* of society, in which economic man would be able to carry out his activities free from all restrictions. This was bound to lead to the best possible results, for, according to the laws of economics, the pursuit of self-interest, in any way short of actual crime, could only result in the maximum satisfaction for all. There was no need for legislative interference, indeed it was nearly always harmful, for was not man 'led by an *invisible hand* to promote an end which was no part of his intention'?

Laisser-faire economics, for Adam Smith and his followers, was the *natural order* which replaced the providence of God or the wisdom of princes. In spite of its limitations, of which we are now only too well aware, and of the terrible consequences that have resulted from it, this doctrine was in its time a great and liberating one. And Adam Smith did far more than announce it. To prove it he laid the foundations of logical method in economic thought, which has outlasted the conclusions that he himself drew from it.

THE LABOUR THEORY OF VALUE

The most important of these foundations is his treatment of the *value* of a product, which he puts down squarely to the *labour* involved in making it. At the time it was enunciated, the *labour theory of value* was directed against the courtiers, churchmen, and the landlords who claimed rewards for which they had not worked. Later, in the nineteenth century, it fell into disrepute among orthodox economists because it was felt that it might be used in the same way against capitalists. In fact the idea of *surplus value*, which Marx showed to be the difference between the value of a product and the wages paid to the actual producers, is already implicit in Adam Smith. The general objective of the search for *profit*, though it is the motive of all economic activity, may, in his opinion, not always be a force for good. The high priest of capitalism has indeed little faith in the probity of his flock, for he calls them

an order of men, whose interest is never exactly the same with that of the public; who have generally an interest to deceive and even to oppress the public, and who accordingly have, upon many occasions, both deceived and oppressed it.[7.149]

The effect of the *Wealth of Nations* was as lasting as it was immediate. It affected far more than economic thought; it provided intellectual justification for the practice of industrial capitalism and gave the tone of a new attitude to a society in which *enlightened self-interest* became the crowning virtue. It was to become a main bastion of liberal philosophy, and from that time on no other such substantial intellectual support was to be given to it.

In other countries, not quite so well situated, economic studies did not lead to such simple and comforting solutions. The earlier American revolution had been an even more frankly economic affair. Its great theoretician, Franklin, had, as we have seen (pp. 526 f.), combined his scientific, economic, and political knowledge with a very concrete practice as a printer, as a postmaster, as a soldier, and as a statesman. The redress of the basic grievance of the colonists, 'no taxation without representation', could be achieved only by active rebellion. Even after its success economic conflicts convulsed the new republic.[7.8]

315. During the American War of Independence (see pages 529 and 805) armoured ships – monitors – were used in sea battles. A woodcut from the *Illustrated London News* of 1862 from a sketch by their special artist at Memphis.

THE PHYSIOCRATS

The pre-revolutionary movement of the *physiocrats* in France was largely an economic one. They based themselves in part on an analysis of England's prosperity and wished to see French natural resources, particularly the land, put to a rational use, which inevitably placed them in opposition to a feudal order which had long outlived its usefulness. Their doctrines, which were also well suited to the rising French manufacturers, had an important effect in guiding the policies of the first French revolutionaries. It was, in the words of Quesnay, a revolution in pursuit of '*laisser-faire*', and its onset was determined by an economic crisis. *Laisser-faire* and *laisser-passer* were in the first place the slogans of a movement to secure manufactures free from government controls and a commerce free from vexatious local tolls and taxes.

THE NOBLE SAVAGE

Eighteenth-century revolutionary social science was not confined to economics. With it also came a new interest, the study of man, not so much as he was in the actual civilized countries of western Europe, but uncorrupted in the savage state. This highly idealized picture had been

316. 'The Inside View of a House in the Island of Ulietea with an Elegant Representation of a Dance to the Music of that Country.' The 'Noble Savage' as depicted in *A New Authentic, and Complete Collection of Voyages Round the World*, London, *c*. 1780, which was concerned with the four voyages of Captain James Cook (1728–79).

built up from the great voyages of that time and by the stories of the missionaries. In addition the literati of Europe were much impressed by what was being learned, also largely from missionaries, about the older and most polite civilizations of China and India.

For the first time since the Greeks there was the possibility of a comparative sociology, and that comparison was by no means favourable to the civilization of eighteenth-century Europe. The more classical of the philosophers, such as Montesquieu (1689–1755) and Voltaire (p. 514), used the comparison with the savages and the oriental civilizations to point their criticism against their own institutions.[7.178] They were concerned, as indeed were all the philosophers and economists, with a great political struggle against the *ancien régime* or, in Britain, against the landed interest; that is, in both cases, against the modified relics of a more or less feudal order. For this they were obliged to appeal away from tradition, from which that order drew its justification, to a hypothetical rule of reason, in which everything was for the best because the laws of Nature could operate without interference.

The *philosophes*, however, had no desire to go to excess. A removal of abuses without upsetting the social order would be quite enough. The romantics went further.

ROUSSEAU

The doctrine of Rousseau, the son of a Genevan craftsman and deeply influenced by the Calvinist tradition, was that the natural virtue of man had been depraved by civilization and could be restored by a return to Nature. Here the old doctrine of the fall of man was given a new twist. Man need not wait for salvation from on high, he could achieve it by his own efforts. Some of the values of civilization, such as law and order, might, however, still be retained by a 'social contract' freely agreed to by the people. This doctrine was to produce an optimistic and essentially democratic revision of social analysis affecting many of its aspects, particularly education.

THE AGE OF REASON: VOLTAIRE

The social sciences of the eighteenth century were for the most part essentially deductive and critical in character. This was natural, as they had come into existence in the first place as a movement of protest against the traditional views by which the Churches, Protestant and Catholic alike, kept their control of the practices of social life. Their prime object was to destroy the intellectual and moral bases of the traditional picture of society – by philosophic criticism, as with the

317. Copperplate engraving of a mob storming the Tuileries on 20 June 1792, taken from an original by Romain Prieur (?–1879).

sceptics Hume and Bayle (1647–1706); by the assertion of materialist views, as with the French encyclopedists Diderot and Holbach (1723–89); by appealing to Nature, as with Rousseau; or, most effectively, by the plain and biting ridicule of Voltaire, with his slogan *écrasez l'infâme* aimed at the Church as the champion of the old régime. Between them they thoroughly undermined the faith of the polite world, which it must be remembered was a very small fraction of the population. What they put in its place was soon seen to be artificial and arbitrary. In the 'Age of Reason' it was impossible, without a much more searching analysis, to find a workable social system. What all of them lacked was an approach which was historical and practical at the same time.

VICO AND THE 'NEW SCIENCE' OF SOCIETY

One exception, who at least recognized the need for a historical approach, was Giambattista Vico (1668–1744), an obscure professor of law in Naples, who early in the eighteenth century, and quite outside the main stream of scientific and philosophic thought, produced his *Scienzia nuova*,[7.172] the first explicit exposition of a science of society. Vico, reacting against Descartes and much in the spirit of Bacon, whom he greatly admired, tried to understand society not by pure reason, but by the character of its products, and especially by its laws and by its poetry. Vico was the first to say expressly: 'Human society is made by man, therefore man can understand it,' and to see that the literature and laws of past periods reflected characteristically the social development of their times. He asked himself, for instance, why epic poetry was written only at the beginning of the classical period, and concluded that those epics were particularly adapted to the society of barbarian chieftains who ruled in Greece in those early days.

He was the first to see clearly that society in all its manifestations – poetic, legal, religious – was a unity, and that that unity was not a static unity but subject to transformation. The movements of history determined institutions. This was of course entirely out of harmony with the prevailing eighteenth-century view that there was one natural and reasonable order of society. The *philosophes* believed that once the anomaly of governments and existing superstitions – the rule of king and priests – had been removed, man would revert to that natural order and live happily ever afterwards. Vico saw further and recognized a necessary social evolution; but even he could not get beyond the old idea of a preordained cycle of human social development, which he explained in terms of divine providence. Yet his work, though unrecognized and unacceptable at the time, was not entirely wasted. Through Hegel and Michelet it was to have some influence on Marxism.[7.126]

THE FRENCH REVOLUTION: THE 'RIGHTS OF MAN'

If the philosophic criticisms preceding the American and French revolutions were to shake the beliefs of polite society, the revolutions themselves were to carry those criticisms to the whole mass of the people. Tom Paine (1737–1809), fresh from the new America, attacked in his *Rights of Man* the whole conception of the ordered civilization of the eighteenth century from Locke to Burke, challenging it as corrupt and tyrannous.[7.56] The very idea that men as men, and not by virtue of rank and wealth, had rights needing to be respected had been new and shocking even when it was only announced in drawing-rooms. Robert Burns, one of the first men of the people to become a great poet, expressed it well in 'The rank is but the guinea's stamp... A man's a man for a' that'.* The further conclusion, that all other rights were arbitrary

318. A guillotine. It was named after a French physician, M. Joseph Guillotin, who, however, was not the inventor and was not, as is popularly supposed, executed by it. From the *Encyclopaedia Britannica*, third edition, supplementary volume, Edinburgh, 1801. The French Revolution was attacked in England, and mob violence often accompanied those who supported it (see illustration 317).

and had to yield in the last determination to those of the common people, as demonstrated in the streets, was the Revolution itself.

However, the French Revolution, necessarily a popular one in its destructive phase, changed its character when the aims of its original promoters had been achieved. Its bourgeois protagonists found themselves as anxious to vindicate the right of private *property* – including, in America, that of slave-owning – against the *mob* as they had been against the king. Most political thinkers were of their persuasion, though Babeuf (1760-97) in France had the daring to claim the liberation of man from his economic as well as his political chains, and was executed in consequence. Up till then the working classes had been ignored in official sociology, except as convenient units in agricultural production or hands for manufacture. They were in fact, at that time, too weak and unorganized to make any serious bid for power.

In late eighteenth-century England the upper bourgeois were already securely in power, and the Industrial Revolution was enriching them mightily. They had no reason to enlarge the area of liberty, but many for restricting it. Already, however, there were stirrings among the lower ranks of the manufacturers and the craftsmen. The French Revolution, and even the Napoleonic era, with its clean sweep of ancient reaction in Europe, roused a sympathy among the people that needed a firm repression to keep it down. It had a deep effect on intellectuals, both on philosophers like Godwin and poets like Shelley, Blake, and Byron. Even the lost leader Wordsworth had written in his *Prelude*:

> Europe at that time was thrilled with joy
> France standing at the top of golden hours,
> And human nature seeming born again. . . .
> Bliss was it in that dawn to be alive,
> But to be young was very Heaven !

12.6 Utilitarianism and Liberal Reform

THE TRIUMPH OF BOURGEOIS LIBERALISM

The fall of Napoleon and the reaction that followed were to turn this optimism away from politics in the direction of a justification and apology for rising industrial capitalism. On the positive side there was the great school of utilitarians with Jeremy Bentham (p. 545) and James Mill (1773–1836), who aimed at showing that if certain abuses were removed there was no reason why free enterprise should not bring

319. Caricature of sweated labour by George Cruikshank (1792–1878).

'the greatest happiness to the greatest number' (p. 533).* On the negative side there was the parson Malthus, who thought that there were far too many of the greatest number anyhow, and that those who could not become petty capitalists and practise self-restraint were doomed to be periodically cut off by famine, plague, and war, as he explained in his *Essay on the Principle of Population* in 1798. He was the originator of that pessimistic doctrine, the *law of diminishing returns*, which has been used ever since to discourage all attempts to improve the lot of man.

RICARDO

A friend of Malthus, though one who disagreed with him in many respects, was the wealthy banker Ricardo (1772–1823), who in the early nineteenth century did much to complete the work that Adam Smith had begun a generation before. He had the experience of the enormous upsurge of the Industrial Revolution behind him and of the security of living in Britain, the greatest, indeed almost the only, manufacturing country. He was accordingly able to write economics in terms of the assured success of the new capitalist 'machinofacture'. He accepted Adam Smith's labour theory of value, but failed to take account of the changed conditions in which fixed capital played a larger share in determining costs. It was Marx who pointed out that this was not really in contradiction to the labour theory of value, as this capital was only labour embodied in plant and machinery.[7.106]

Ricardo's failure, however, was made the excuse by later economists to upset the whole labour theory of value, and turn economics away from an interest in the production of goods to one in their exchange. Indeed this process was already beginning with Ricardo's own work. As a banker he was more concerned with the distribution of wealth than its creation. This led him to consider the relative proportions of *rent*, *interest*, and *profits* into which the surplus of value less *wages* was divided up. Interest and profit were justifiable, but as, in his view, land in itself could have no value it was difficult to justify *rent*. To do this he had to introduce the idea of a *differential* rent of land better than the average. This idea, inconsistent with his other views, was to be later extended to profit as well. It gave rise to the differential or *marginal* theories of value, which, as we shall see, were introduced later to conceal the *exploitation* implicit in the labour theory of value. Ricardo's main objective was still to discredit the landed interest with their policy of protection and dear corn. His work accordingly contributed to the ideology of the reform movement and led to the triumph of free trade. He was not, however, as were the later economists – the 'vulgar economists' of Marx – Say (1767–1832), Nassau Senior (1790–1864), and Bastiat (1801–50), obliged to justify or apologize for capitalism 'with bad conscience and evil intent', under the first stings of the criticism of the lower orders.

THE IRON LAWS OF ECONOMICS

Towards the mid nineteenth century the social sciences, securely based on utilitarian logic and justified by business success, seemed to be settled for all time. Liberalism and free trade had won the day and 'progress' along the same lines seemed assured. Nevertheless there was no denying that an uncomfortable feeling was growing that there was something very wrong with the whole system. The 'dismal science' of economics palpably aimed at justifying everything as it was, and justifying it at a time when to the poorer classes, indeed to anyone other than a successful manufacturer, it appeared that something was very much wrong. In the wealthiest industrial country of the world, in the most flourishing commercial centres that were the pride of the age, there was famine and pestilence,[7.48] and everywhere ignorance and social insecurity were multiplied on a scale that had never existed before. All this had to be explained and excused; no room must be left for any

320. The evils of spirit drinking as depicted by William Hogarth (1697-1764) in his famous engraving 'Gin Lane'. Oblivion could be bought although it generated worse poverty, as Hogarth depicts.

human feelings; pitiless and inevitable law must take their place. The economics, the logic, and the psychology of the mid nineteenth century were accordingly constructed in the driest imitation of physical science, and were as far removed from living actuality as they could be.

RELIGION AND REACTION

Even religion had to be called in to provide supernatural sanctions for capitalism when rational ones failed to convince. There had been after the French Revolution a deliberate return to religion to counteract the eighteenth-century scepticism of deism,[7.65−6] the spread of which had shown itself so dangerous to law and order. The wealthy returned to church-going and some even toyed with Roman ritualism. Methodist and other evangelical movements satisfied the emotions and did something to divert the criticism of the lower middle class and the deserving poor, though the 'Nonconformist conscience' was often found supporting the radical side. But religion was not allowed seriously to interfere with the solid common sense or the conduct of business of the middle class. It was not so much that, according to the unbeliever's jibe, religion was kept strictly for Sundays, it was rather that scriptural

321. 'The Factory Children' from *The Costume of Yorkshire*, Leeds, 1885. The originals for this work were prepared in 1814 by George Walker (1781–1856). The picture shows two stunted children in front of the mill where they are employed.

justification could be found for abstinence and thrift as well as for the material rewards which came from practising them. Though this was the dominant and official note, by no means all the voices that were raised were in favour of acceptance and resignation.

RADICAL AND ROMANTIC PROTESTS

Indeed poets, writers, and artists protested in words and pictures against the new ugliness of nineteenth-century capitalism and, in looking at that ugliness, were brought to see the horror and misery that underlay it and the smugness and greediness that perpetuated it. In the revolutionary period Blake, Byron, and Shelley had lifted their voices against tyranny and grinding poverty. Later, when capitalism seemed to be established for good, novelists like Dickens, Mrs Gaskell, George Eliot, and even Benjamin Disraeli attacked its manifestations mercilessly, but had no alternatives to offer. Critics like Carlyle and Ruskin tried to lay bare its origins and to supply remedies; as, however, they had neither the wish nor the understanding to push their analysis to the point of the class and property basis of society, they tended to find their solution by turning to the past, to the age of heroes or of faith, to anything but the commercialism and industrialism of their day.

ROBERT OWEN

More effective, because he came closer to the people, was the work of the Welsh radical, Robert Owen (1771–1858). He was characterized by a reluctant admirer[7.68] as a man 'who made a fortune by capitalism and common sense and squandered it in communism and craziness'. A self-made man, he rose to be a wealthy manufacturer by the time he was twenty, and showed practically in his New Lanark mills, into which he introduced education and social welfare, that the slavery and squalor of the cotton trade were not even good business.

Robert Owen had a constructive and economic mind rather than a political one. He was one of the first to recognize the ability of the new machinery, if properly run, to provide wealth for all. However, in common with many liberal idealists, he believed that all that was necessary for reform was to prove to the wealthy and powerful that it would be to their advantage to accept and even promote a juster social system. His attempts to secure action from the workers themselves were always limited by his fears of alarming the ruling classes. His aim was a co-operative commonwealth, what he called a New Moral World, which was to organize production without distinction between capitalists and workers. His first attempt was to set it up in the free New World, but his Colony of New Harmony in Indiana in 1825 lasted barely three years and its failure absorbed most of his fortune. Undeterred, he returned to England, and began to put his remaining money and drive behind the Trade Union movement, just then emerging from illegality. In 1833 he had turned the Builders' Union into a Builders' Guild aimed at eliminating the contractor. Next year he became even more ambitious and launched the Grand National Consolidated Trades Union, which was broken in a few months by the combined attacks of government and owners. In the end the co-operative venture proved the most lasting. Owen himself favoured producers' co-operatives, but it was in distribution that it was to strike root. The first enduring co-operative shop, though by no means the first to be set up, was that of the Rochdale Pioneers in 1844. In spite of its material success and the degree in which it protected the poor from excessive prices, it could do nothing to stop the fundamental exploitation of the profit system, nor even at

322. By the midnineteenth century some attempts were made to help the working man educate himself. In 1859 the Warrington Mechanics Institution introduced a mobile library with the result that book circulation rose from 3,000 to 12,000 books per annum.
c

its greatest extent, in the present century, have the co-operative societies provided for more than ten per cent of retail trade.

The co-operative movement, however mellowed with time, remains as an example of the ability of the working class to form and run their own organizations. The deepest understanding of the social consequences of the Industrial Revolution was to come from those who were its victims – the radicals and Chartists – coming mostly from the ranks of the small traders and craftsmen. Men like John Gray (1799–1850) and Francis Bray (1809–95) carried the arguments of the utilitarians and of Owen to the logical conclusion of a socialist state in which all will be producers of wealth. Another was Thomas Hodgskin (1783–1869), the real founder of the London Mechanics Institute, later Birkbeck College. His criticism cut much deeper, for he found the only alternative to the permanent impoverishment of the working class in the abolition of the capitalist (p. 1150).[7.9; 7.28; 7.68]

These men were not content to analyse and criticize society; they were trying to change it, to achieve political and economic liberty, to secure the legality of trade unions, and to transform the limited advance of the Reform Bill by the popular democracy of the Charter. By the middle of the century it appeared that their struggles and sufferings were in vain. The capitalists were more firmly in the saddle than ever. Some injustices were, it is true, mitigated – some share of the new commercial prosperity fell to the workers – but the basic injustices of exploitation remained. Nevertheless the British forerunners of socialism, however Utopian their aims, brought to the movement a wealth of experience and enthusiasm which was to be of permanent value.

FRENCH SOCIALISM

The workers in France, for all its revolutionary tradition, were no better off than those of Britain. As the immediate effect of clerical reaction disappeared it was replaced by a cynical domination of wealth, imperishably portrayed by Balzac and Daumier. The Revolution of 1830 brought to the French people as little liberation as the Reform Bill of 1832 brought to their English brothers. Nevertheless, the tradition of revolution in France in those days still stimulated a lively and intelligent discussion of social and economic questions. The half-crazed aristocrat Saint-Simon, and the practically-minded François Fourier, the founder of the European co-operative movement, argued more abstractly and more systematically than the British philanthropists and radicals. They determined, once and for all, the character and organization of a new society embodying the ideals of social justice and freedom.

They believed that once people understood that the new form of society – Socialism (for the word dates from the 1830s) – was infinitely superior to the old, they would naturally acclaim it and that after a revolution, preferably peaceful, the millennium would arrive.

It did not happen that way; Owenism, Chartism, or Utopian socialism, for all the enthusiasm they generated, failed to effect any significant change in the domination of capital. It was clear that they all lacked something essential: an understanding of the workings of society sufficient to enable the clearly desirable changes to be brought about. Technology had found the answer to its problems by building up the physical sciences, for, as we have seen, those sciences grew when and only when it was possible to use their discoveries of the workings of Nature to control it. What was needed in the nineteenth century was a *science of society* that would be effective in the same way in securing the control of society by the people who formed part of it.

12.7 Marxism and the Science of Society

To create that science of society was to be the great achievement of Karl Marx and Friedrich Engels. Karl Marx was born at Trier in the Rhineland in 1818, the son of a liberal-minded and cultured lawyer. Both time and place were propitious for one who was to bring about such a radical change in human thought. Though close enough to France to feel the whole effect of the great intellectual movement of the Enlightenment, he was brought up just outside the main stream of capitalist development. He thus escaped the danger, common to both French and British thinkers, of taking it for granted. But these advantages would have been of no avail if he had not grown up with a combination of keen intellectual insight and a deep passion for human justice.

His student years were spent at a time when intellectual and political discussion was at its height in the general ferment which preceded the Revolution of 1848. Germany had not yet participated in the Industrial Revolution and had, through the Napoleonic wars, come under the influence of French liberal thought. The result was that in Germany it was possible to think abstractly about social matters without being incommoded by too close a comparison with practice. The late eighteenth and early nineteenth centuries were the great periods of the German idealist philosophy of Kant, of Goethe, of Schelling, and, most of all,

323. Karl Marx (1818–83) who with Friedrich Engels created the science of society.

of Hegel. In Germany the social and natural sciences were combined in one grand *naturphilosophie* (pp. 645 f.). Kant did not think it incongruous to propose the nebular hypothesis for a solar system and the categorical imperative for human behaviour on Earth; Goethe sang of ideal beauty and studied the cervical vertebrae of the mammalia; Hegel produced a philosophical system which included everything from the absolute to the ideal Kingdom of Prussia, which he deduced by an idealist dialectic method, starting from nothing more than the idea of existence entailing the idea of non-existence (pp. 182 f.).

The logic of Hegel had nevertheless certain advantages. It was more fluid and imaginative than that of the eighteenth-century thinkers and it included a factor they nearly all lacked: a sense of historical development. It forged the necessary tools for studying situations that Hegel himself did not care to face. After Hegel's death in 1831 his followers tended to divide into two camps, the right Hegelians, who emphasized the idealist and conforming side of his philosophy, and the left Hegelians, who developed his ideas of dialectic change in a revolutionary sense. An influential Hegelian who turned from idealism to materialism was Ludwig Feuerbach (1804–72), who attacked the corner-stone of legitimist reaction – institutional religion. Feuerbach, unlike the eighteenth-century sceptics, did not call religion a conscious fraud, but he did say it was itself a creation of man in society. For him the Holy Family in heaven was but an image of the human family on earth.

The young Marx found the ideas of the Hegelians stimulating but inadequate. He felt that Feuerbach had only begun his task in exposing the inner nature of society and religion. 'The philosophers have *interpreted* the world in different ways; the point, however, is to *change* it.'[7.111.473] And now new opportunities for this were opening up.

Already, in the early forties, revolution was in the air; all over Europe the repressive system of the Holy Alliance was being challenged. Marx, starting his active career as a liberal journalist, found that he had to reckon with social and economic realities and learn his philosophy from the actual rather than the ideal world. He had no doubts as to which side he was on. He wrote articles against the landlords and in favour of oppressed peasants.[7.125] Finally he came to grief for criticizing the Prussian censorship and had to flee from German soil. He only returned for a short time to take part in the unsuccessful revolution of 1848. Through his exile in Paris from 1843 to 1845 he came into contact with French revolutionary and socialist thought. Even more important was his meeting with his young compatriot Friedrich Engels, who had lived and worked in Manchester and had seen the achievements and the horrors of the Industrial Revolution, and the upsurge of the Owenite and Chartist movements.[7.9; 7.28]

HISTORICAL MATERIALISM

Marx was at the point where all these influences converged, and his combination of clear understanding and passionate concern for social justice enabled him to fuse them together into a coherent and essentially new theory of society. He brought Hegelian logic, imbued as it was with the ideas of change induced by inner conflict, to bear on the struggle of the classes of his own time. He was thus enabled to explain the inner dynamic of all social movements. He provided the missing factor of all the previous socialist analyses by showing that the motive power of social change was neither providence, as the religious asserted, nor, as the liberals maintained, the pursuit by reasonable people of an ideal state of affairs. He found that motive power in the struggle which oppressed, but rising, classes waged to secure a tolerable and fuller life. He showed further that the new classes had arisen one after the other on account of technical and economic changes in methods of production and of the changed legal, social, and economic relations between people which flowed from them. At the same time he understood well enough that indignation and a just cause were by themselves no guarantee of success in any class struggle. They needed to be supported by a 'class consciousness'[7.111.204] and by a theory not derived *a priori* but built out of the facts of social history. Before Marx socialist ideology had consisted of mere exhortations or descriptions. He made it a *science*.

PHILOSOPHY AND POLITICAL ACTION: THE 'COMMUNIST MANIFESTO'

Social science no less than physical science requires action to test it and give it full contact with reality. Marx saw that the important field of action in his time was that of revolutionary politics. He was not content to work out the theories of social change; he understood that the actual achievement of change was the result of thousands of human wills, acting the more effectively the more clearly they understood the science of social change, and the more closely they organized themselves to achieve their ends. He made it his life's work to understand, to explain, and to organize the transformation of capitalist to communist society. The first step was the publication of the *Communist Manifesto* in 1848. This was the first call to action of the new scientific socialist movement. It still remains the shortest and clearest exposition of Marx's analysis of society and social change.

324. A soup kitchen in the East End of London, 1867. Every day, except Sunday, between 3,000 and 4,000 gallons of soup were sold to the poor at a penny a quart. From the *Illustrated London News*, 1867.

What Marx and Engels first brought out in the *Communist Manifesto* was that the agent in the transformation of capitalist into communist society was the working class: the propertyless proletariat, itself brought into being by the rise of the capitalist method of production for profit. Now the reputable historians and philosophers of the period had long been accustomed to class divisions in society, and the experience of the French Revolution had brought them vividly to their attention. But it had never occurred to them, and it could never occur to them on account of their own upbringing, that the working or lower classes should have any other function in society than assuring the well-being of their betters. To most of them they were a mob needing to be kept down.[7.183] The more tender-hearted considered that at least some of them, the deserving poor, were suitable objects for charity, while liberals felt that they ought to have a mutual interest with their betters in maintaining the existing state of society.

THE SPLIT IN SOCIAL SCIENCE: BOURGEOIS AND MARXIST

The picture of the social history of western Europe in terms of class struggle was so foreign to academic thought, and so unlikely to appeal to the powers controlling universities, that it is not surprising that it has taken more than a hundred years for it to penetrate into the official social sciences. Even today in the countries of 'freedom' and 'democracy' Marxism is only officially taught, in the very few places where it is taught at all, by anti-Marxists, though it must be admitted that a great deal of the efforts of the official economists and sociologists is now taken up with denouncing the errors of Marxism.

From the time of the *Communist Manifesto* there have been effectively two distinct and rival trends in social thought. While there is no precisely formulated bourgeois social science, all the varieties of social theory, from the rugged individualism preached in the United States to the idealism of the social democrats, are united on a common basis of social assumption which makes them, even if controverted in detail, essentially acceptable to the powers that be. And rightly so, for they all concur in effect that capitalism is a most proper basis of society, and that it will last for ever or for a very long time. They are all agreed that, whatever its defects, it is superior to the kind of socialism that they imagine is growing up in Russia and that is now spreading over increasing areas of the world. The acceptance of capitalism, either unaltered or slightly modified in the direction of a Welfare State, marks adherence to general idealistic and unhistorical theories of society. These theories are very far from being scientific, and they are in any case made little use of in directing the affairs of the capitalist world.

PRODUCTIVE FORCES AND RELATIONS

Marxist social science is, in contrast, first and foremost materialistic and historical, and is proving its worth in full-scale social experimentation. It takes each state of social organization, including capitalism, as simply one historic stage in a succession of economic phases of the utilization of material natural resources for man's material needs. At any period the form of society is limited primarily by the technical level of production, the *productive forces*, embodied in the actual material means of production. A mass-production factory, for instance, requires a far more elaborate organization of society than does the hunting of kangaroos. The full use of the means of production depends on the existence of *productive relations* – of buyer and seller, master and man – appropriate to the means of production at the particular time.

In actual practice there has always been a time lag between the available *productive forces* and the contemporary *productive relations*, which are for the most part those belonging to the means of production of an earlier phase. This time lag is a reflection of a *conflict* between the *classes* of the society, which is the essential motive power for social change. The victory of the class which can bring about productive relations more in harmony with the contemporary productive forces leads to a higher stage in society, and also to a rapid further improvement in the means of production. That improvement, which has been immensely accelerated by science in the last two centuries, has been a factor in creating social instability, but cannot in itself be a cause of social change, which must always have a human motive force.

The productive forces and the productive relations do not in the Marxist view cover the whole of social reality. Each new class, in the very struggle to establish itself, comes into conflict with the ideas, styles, customs, and laws of the old ruling class – all part of the *ideological superstructure* – the means by which it maintains its dominance. To secure its victory, the new class must forge a new and different ideology that gives it the knowledge and inspiration to achieve power, and that, once that power is secured, becomes the dominant ideology of the new stage of social development. Thus, as we have seen, the individualist ideology, growing up in the latter stages of feudalism, helped the achievement of bourgeois power, and became the corner-stone of the liberalism in the name of which it defended the privileged position of property.

CAPITAL

The evolution of society in this way is seen, in the light of Marxism, as a *process* of revolutionary change, and consequently no stage in it, and least of all *capitalism*, is a *state* of society. The detailed justification for Marx's general social theory, after a preliminary outline in the *Communist Manifesto* and in *Wages, Prices, and Profit*, was reserved for Marx's most definitive and classical work – *Das Kapital (Capital)*. The first volume appeared in 1867 and the last two, only in an incomplete form, in 1885 and 1894,[7.106] after his death. In it Marx not only viewed economics historically, showing the contingent and temporary nature of its laws, but also interpreted history in terms of economics. He showed how its major transformations, from the classical slave empires to feudalism and in turn to capitalism, were essentially economically motivated and depended on changes in the mode of production. He thus provided for the first time the logical consequences of cause and effect which were missing from all academic writing of history. *Capital* was based firmly on a scientific analysis of the past, so that it could deal with the working of contemporary economics while fully understanding its impermanence.

325 a, b. Two contrasting pictures of social conditions in London in the 1870s. A poor street trader and his family in the Houndsditch area of London's East End, and a fashionable croquet party. From *London. A Pilgrimage*, by Gustave Doré (1833-83) and Blanchard Jerrold (1826-84), London, 1872.

THE THEORY OF SURPLUS VALUE

Marx took over the labour theory of value, elaborated by Adam Smith and Ricardo, and drew from it the all-important concept of *surplus value*. The 'vulgar economists' (p. 1061) saw the wages paid to workers as a fair exchange for the value of the work they did, and attributed the profit of the employer to his superior foresight or abstention from spending. Marx, on the contrary, took the 'exchange value' of the commodity as the 'socially necessary labour time' required to produce it, unlike the proponents of the marginal theory, who derived it from the subjective estimates of the market itself (p. 1061). It included the value (embodied labour) of the raw materials and of the part of the capital equipment used up in its production (*Capital*, Vol. 1, Ch. 7).

What is paid to the free worker under capitalism is by no means the 'socially necessary labour time' expended. Labour itself is treated as a commodity like any other, and its value is reckoned at what value, or labour time, is embodied in it. It represented just enough for the worker and his family to live on in the customary way for that kind of worker at the time (Marx's historic and moral element). Actual wages might be above or below this value, but tended to coincide with it.

The difference between the exchange value added in production and the value of the labour expended in producing it is the *surplus value* taken by the capitalist, and is the essential source of his wealth which he can use as capital for further operations. It accrues to him not on account of any particular merit or service on his part. He could extract it thanks to the power of the social system of capitalism, established as the result of political and economic struggles by the bourgeoisie and maintained with all the apparatus of coercion: the law and police of the State which they controlled. Marx thus demonstrated that capitalism, far from being a system of natural justice as the classical economists maintained, was one of extortion imposed by arbitrary force. It was this central concept of *exploitation*, which the economists had hidden even from themselves, that was to give explosive force to Marx's teaching. It was not that workers and peasants had not felt exploited for centuries, it was rather that now they could understand the mechanism through which the exploitation was carried on. They could see that it was not a matter of the hardness of this employer or that landlord, but that the fault lay with the whole system by which the employer and landlord lived. The workers were also to learn with Marx and his successors how to 'expropriate the expropriators', and how to build for themselves, in harmony with the laws of social development, an economic system in which wealth produced socially should be socially distributed.

Marx's social and economic theory, set out in all essentials by 1867, was in the direct line of the evolution of the study of society from myth through ordered description to science. It was and it remained as the live and developed branch of social knowledge and action. Yet by ignoring or repudiating it bourgeois social sciences were to continue and even to evolve in serving their purpose as an apology for a capitalism that had passed its zenith but was still far from visible decay. The implications of Marxism were so repugnant to the ruling bourgeoisie of the time that it is not surprising that for another fifty years all over the world, and until today in a large part of it, they were ignored and rejected by the official exponents of social science.

TWO SYSTEMS OF SOCIAL SCIENCE

It will accordingly be necessary in the remainder of this chapter to treat separately the progress of the two intrinsically incompatible approaches to the science of society, represented by the work of the bourgeois sociologists and the Marxists respectively. They have followed in the last hundred years divergent paths. For many years the bourgeois social scientists ignored Marxism from mere unawareness of its existence. Even when the progress of Marxist socialism in Europe could no longer be hidden, the learned world continued to ignore it on principle or to dismiss it as beyond the pale of science. Until 1917 it had no official status: it belonged to an underworld of subversive propaganda. As Marx referred to it half-jokingly in 1856:

> We know that to work well the new-fangled forces of society, they only want to be mastered by new-fangled men – and such are the working men. They are as much the invention of modern times as machinery itself. In the signs that bewilder the middle class, the aristocracy and the prophets of regression, we do recognize our brave friend, Robin Goodfellow, the old mole, that can work in the earth so fast, that worthy pioneer – the revolution.[7.112.428]

In the Britain where Marx and Engels lived and worked, their ideas had the least influence on the self-satisfied bourgeois intellectuals; but even there, as will be shown, the influence through William Morris and a group of fellow Marxists in the Social Democratic Federation was serious enough to need some refuting. In Europe, especially in Germany and France, it was much greater, and the uncomfortable awareness of it spread far and wide and even influenced in one direction or another the majority of social scientists. And while they refused to acknowledge it, academic social scientists could not fail to pick up and even assimilate many Marxist interpretations, though they might repudiate the methods of Marxism and the practical conclusions drawn from them. In particular, the use of historical method and more or less distorted forms of the economic interpretation of history became widespread and respectable.

In the light of history we can now see something of the magnitude of the contribution of Marx and Engels in the creation of a new science of society. It was an achievement intellectually comparable to that of Galileo in the physical sciences or of Darwin in biology. But just because it was an achievement which touched every aspect of human life even more closely, just because it roused at the same time every interest vested in the old order of society and every aspiration of its victims, it was to be in effect far more important than the greatest discovery in any field of natural science. Because it meant much more it had to fight far harder for recognition. Knowledge of Marxism is essential to the understanding of the place of science in history. Without Marxism natural science would have remained as a growing accumulation of interesting facts about the universe and useful recipes for controlling it; human history would still be restricted to the simple narration of political changes without any coherent thread of explanation.

12.8 Academic Social Sciences
in the Late Nineteenth and Early Twentieth Centuries

The main body of academic thought was to continue along already classical liberal lines in the latter part of the nineteenth and early twentieth centuries, though a strong counter current of mystical and irrational thought was to appear towards the end of the period. A much more pervasive influence came from the biological sciences, particularly the newly enunciated theory of evolution, though psychology was also influenced by physiology, and archaeology by palaeontology. The association of the social and the biological sciences was to bring to the social sciences something of the habits of observation and *inductive* logic, and thus to break down to some extent the habit of using *deductive* argument from first principles that they had inherited from Aristotle and the Church. But it was also to produce a belief that sociology was simply human biology, which was to have such catastrophic results in our own time.

The common characteristic of all late nineteenth-century social sciences, in contrast to those of the eighteenth century, was their evasiveness. In this they merely echoed on the scientific plane the prevailing hypocrisy of polite society. Just as there were aspects of life known to exist and never mentioned, so in the social sciences references to the existence of classes or of exploitation were beyond the pale. Explanations of social phenomena had to be concocted which made no allusions to these awkward facts, and if that could not be done it was better to have no explanations at all.

THE EXTENDED FIELD OF HISTORY:
THE DISCOVERY OF EARLY CIVILIZATION

The most inclusive of all social sciences, history, underwent no important theoretical changes in the period under review. Yet history was in that period, especially through its ancillary discipline of archaeology,

326. It was towards the middle of the last century that archaeological excavations began in earnest with the work of Austen Layard (1817–94). His work at Nineveh and Babylon opened a new era in early historical research. This drawing of the 'Fish God' Dagon and the removal of debris shows the excavation in action. From his *Discoveries in the Ruins of Nineveh and Babylon*, London, 1853.

so extended in range and detail as to provide completely new horizons of time. Piece by piece the story of the 'prehistoric' past was revealed. The old literally interpreted chronology of the Bible was found to be as hopelessly inadequate for the history of man, even of civilized man, as it had been found by the geologists for that of the world of rocks, plants, and animals. The Stone Ages, old and new, were revealed and the historic sequence of their culture established. The literal history of the early civilizations of Egypt and Mesopotamia began to be read from contemporary documents. More was being learned of the history of the other civilizations of India, the Far East, and of the Americas. At the same time hundreds of thousands of detailed facts from documents or excavations were being fitted into the classical, medieval, and modern European histories.

The new historical horizons could now have provided, for the first time, the framework of events through which the present grew from the past. Yet the general interpretations of this new picture were undertaken only by eccentric historians like Buckle or Winwood Reade. The professional historians tended to specialize on limited places and periods, and pride themselves on writing 'scientific' history with the minimum of interpretation. The writing of history had not always been so restricted.[7.165]

PROPAGANDIST AND SCIENTIFIC HISTORY

Originally history had been written for unashamedly propagandist purposes. At first it aimed at glorifying the deeds of the Heroes and Kings, Cities and Churches. Later, all through the controversies of the Renaissance down to the early nineteenth century, it was of political service to one or the other side. The limited objectivity which political and religious history managed to reach was derived, in fact, from real political battles such as those of the Reformation and counter-Reformation or of the Whigs and the Tories. The weak points of the antecedents of either side were well brought out by their opponents. Even the greatest of the eighteenth-century histories, Gibbon's *Decline and Fall of the Roman Empire*, was essentially a tract for the times, an attack on the Church as the agent of corruption and decay.

This tendency continued well into Victorian times, as the histories of Macaulay, justifying Whiggery, or of Froude, idealizing Elizabethan imperialism, show. However, largely due to German influence, a tendency to objective history spread. Indeed a historian was considered the more scientific the less he attempted to explain why things happened. A reputation for objectivity could best be built up by sticking to one's own period and avoiding generalizations.[7.165] But this was a very one-sided and deceptive method. By collecting mere facts and by refusing to have theories about history, historians of the period were justifying by default the existence of the system under which they lived. It was an episodic and meaningless history corresponding to an individualist and unregulated economic system. The fundamental scepticism of the official historian has been expressed in H. A. L. Fisher's Preface to his *History of Europe* :[7.57]

Men wiser and more learned than I have discerned in history a plot, a rhythm, a predetermined pattern. These harmonies are concealed from me. I can see only one emergency following upon another as wave follows upon wave, only one great fact with respect to which, since it is unique, there can be no generalizations, only one safe rule for the historian: that he should recognize in the development of human destinies the play of the contingent and the unforeseen. This is not a doctrine of cynicism and despair. The fact of progress is written plain and large on the page of history; but progress is not a law of Nature. The ground gained by one generation may be lost by the next. The thoughts of men may flow into the channels which lead to disaster and barbarism.

The concealed reason underlying such an attitude was that any serious and rational attempt to interpret history would be bound to lead to criticism of the existing economic system or, even worse, to Marxism. The most that history was expected to show was 'progress', and even that, as the century drew to an end, became more doubtful.

An apparently opposite view was that which contrasted the idea of history as a science with that of history as an art, a vehicle for fine writing and amusing stories. This gave ever more scope for romantic reconstructions of the past as official histories became more objective and dull. As a result the gap of historical interpretation which the serious historians failed to fill was left to propagandists of nationalism and imperialism, to ignorant fanatics, to downright reactionaries with race theories and prophetic fulfilments.

THE FOUNDATION OF ANTHROPOLOGY: MORGAN AND TYLOR

What history failed to do – bringing men to understand the society in which they lived by looking at its past – might have been achieved by the more immediate study of the variety of social patterns of peoples with very different levels of culture living in the contemporary world. The new science of anthropology had indeed a promising start in the mid nineteenth century, when L. H. Morgan's (1818–81) and E. B. Tylor's (1832–1917) studies of American Indian and other cultures revealed the

327. Romantic reconstructions of the past were in vogue around the sixties and seventies of the last century. Francis Topham (1808–77) was an artist whose 'feeling for picturesqueness' and romanticism show well in his painting 'A Roman Triumph', now in the City Art Gallery, Leicester.

common social structure of tribes in many parts of the world, with its complex relation system, often traced through mothers, its absence of private property, prisons, or police.[7.116; 7.167] Such, they conjectured, were the more primitive states of our own civilization and they showed how they corresponded to the social organization of Greece or Rome. But this track was left to Engels and later Marxists to develop;[7.49] it was too dangerous for academic anthropologists and for the missionaries and traders who were their chief informants from the field, for it struck at the bases of government, morality, and property. Instead of such a tendentious historical approach it was much safer to follow the comparative methods of Frazer (1854–1941) or Westermarck (1862–1939), to collect objects of art and folk-lore, and to search for racial origins by measuring skulls.

With improved communications and the renewed urge for imperial exploitation that characterized the later nineteenth century, contacts with primitive people were multiplied. Although most of these led to their exploitation or extermination, there was as well far greater opportunity to learn about their customs and beliefs. The first serious field anthropological studies were those carried out by Miklucho-Maklai (1846–88) to New Guinea in 1871 and by the zoological expedition to the Torres Straits and New Guinea in 1898–9 in which A. C. Haddon (1855–1940) and W. H. R. Rivers (1864–1922) took part. But here again these direct observations, while confirming the pattern of tribal organization of Morgan and Tylor, were not pushed beyond the provision of psychological interpretations, and their economic aspects were overlooked.

SOCIOLOGY

The same insistence on accumulating simple and disconnected facts and on the comparative method, together with the rejection of historic and economic interpretation, doomed from its very birth the mid nineteenth-century science of sociology. This subject had been doubly unfortunate in its founders, Comte and Herbert Spencer. Both were the oddest and most unsociable characters imaginable. Comte (1798–1857), a disciple of Saint-Simon, was a recluse who early in life acquired a burning conviction that he held the key to the ideal arrangement of society. This was the application of what he called the *positive* method of science, a method which was destined to supersede the earlier methods of religion and philosophy, and which he expounded in a series of long books.[7.31] In the sense that it broke with traditional or religious explanation this was an advance for the time, but Comte was narrow and dogmatic and essentially reactionary.[7.67] Indeed he had little to add to the science of society but misapplications of the ideas of physical science. He believed that the superior men should lead the people, and he had a corresponding petty-bourgeois dislike of popular movements. Though he acquired disciples, some as intelligent as Harriet Martineau, George Eliot, and John Stuart Mill, and even founded a religion – the Church of Humanity – his work gave rise to little of enduring value. The term *positivism* was to be taken up by Mach later in the nineteenth century in an even more abstract sense.

Herbert Spencer (1820-1903) started life as an employee of the new railways and was a fervent admirer of *laisser-faire* capitalism. Assured of a small competence, he devoted himself to the self-imposed task of charting the development of human society. He had developed, some years before Darwin, the idea of universal evolution.[7.152] Unlike Darwin's, it was based on a hypothetical law of the necessary continuous increase in complexity and differentiation in the world, rather than on any insight into the material mechanism which accounted for it. Spencer reformulated Comte's *sociology* along biological lines. It was his work, which had in its time an immense reputation in Britain and even more in the United States, that did most to promote the idea of sociology as a branch of biology.

THE INFLUENCE OF DARWINISM:
EVOLUTION: BIOLOGISM

In all the fields of social studies in the late nineteenth century; the intellectual justification for the evasion of historical description and of the drawing of contemporary implications was found by reference to evolutionary biology, relying on the immense prestige of Darwin. Though this had originally a liberating effect, destroying as it did the idea of permanently ordained states of society, it introduced other errors which were to help to confuse and sterilize all social studies. Just as, in the early seventeenth century, it was the physics of Newton with its natural mathematical law that dominated social thinking, so in the late nineteenth century it was the biological law of evolution that could be called on to explain all changes.

It was not the first time that biology had been invoked in social theory. Especially in the seventeenth and eighteenth centuries the order of Nature was praised as an example of God's handiwork, and held up to men to guide their own social life.[7.178; 7.179] But the living world was then thought to be as ordered and static as that of the heavens. Early biology was used to defend the picture of an order of Nature created once and for all by God. Paley's (1743–1805) celebrated argument of the watch was almost the last form of this justificatory biology.

With evolution, matters were very different. Evolution was not a static thing, it was a process that was going on all the time, one that was changing the world, and changing it in a way that the men of the nineteenth century could well understand. It spelt *competition* and led to *progress*. What was not understood, or at least not admitted in the nineteenth century, except by anarchists like Kropotkin (1842–1921) or eccentrics like Samuel Butler (1835–1902), was that in evolution men were simply seeing the social practices of capitalism in relation to human beings reflected back to them as a scientific theory about animals and plants in the state of Nature (pp. 644 f.).

GALTON AND EUGENICS

By treating man as merely an evolving animal, the development of human society was completely misunderstood to be a moral evolution as spontaneous and uncontrolled as bodily evolution was supposed to be.[7.40] It was with the highest of intentions that Francis Galton, Darwin's cousin, set about studying the heredity of men of exceptional ability in Britain.[7.64] He found that many were related and that all belonged to relatively few families. Because of his biological pre-occupations he overlooked the historical fact that the ruling class of Britain in his time was a very small minority, very much intermarried, and the social fact that the chances of success, even intellectual success, were, and still are, overwhelmingly weighted in favour of children from cultured and well-placed families (p. 950).

Technically Galton's work marked the first crude use of statistics in inheritance studies, and it led to the foundation of the socio-biological science of *eugenics*, which has ever since concerned itself largely with attempting to prove, on genetic grounds, the superior value of upper-class stocks and stressing the need to protect them against the careless breeding of the inferior poor. This biological interpretation of mankind, with its emphasis on race and breeding, affected to a greater or lesser degree most progressive thinkers in the social and historic sciences. It was popularized by historians like Green and novelists like Wells, who were never able to see that by reducing the development of man to a lower level of evolution they were making nonsense of history and social science.

328. Francis Galton (1822–1911) was the first to use statistics in studies of inheritance. His views were coloured by his relationship with the Darwins and in this woodcut of a collection of 'Most Distinguished Fellows' of the Royal Society that appeared in the *Graphic* in 1889, the following are depicted: (from left to right, standing) G. H. Darwin, Galton, Thiselton Dyer, R. H. Scott, William Huggins, (and sitting) Gabriel Stokes, Joseph Hooker, J. J. Sylvester and T. H. Huxley.

RACIAL THEORY

Worse results were to follow in practice. This double transposition of social images into biology and back again was to have a terrible sequel when applied in practice in the twentieth century. More than anything else it sapped the foundations of the older belief, insecurely established in traditional religion, that man belonged to society and that his very individuality could find adequate expression only through society. Imbued with the false biological view of humanity, as a race rather than a community, even the limited sanctions implied in religious morality fell away. Life became a free-for-all where the doctrine of race could be used to justify any degree of class or colonial exploitation: could even be used to prove that white and black men were of different species.[7.39] Its full horror, however, was reserved for our time, where the excuse of race superiority, fanatically believed in by thousands of the followers of the Nazis, was used to perpetrate under conditions of incredible cruelty and degradation the largest and most senseless massacres in history.[7.134] The Nazis probably killed in cold blood more than the sum total of all those massacred by all civil and religious intolerances in the whole of human history; and all in the name of a biological theory.

This is only half the charge; besides glorifying race this perversion of Darwinism also glorified war, for it was in war that the race proved itself and it was in war that the fittest survived.[7.84] It is true that these ideas were for the most part those of ignorant and fanatical men. Though philosophers such as Nietzsche (1844–1900), Bergson (1859–1941), and Sorel (1847–1922) gave them some countenance, very few scientists had any direct part in them. Nevertheless, nineteenth- and twentieth-century scientists cannot escape the blame. Their fear of entangling themselves in politics meant that they left the social application of their own ideas to other people, and made no effective protest against the perversion of the products of their own researches.

ECONOMICS: THE MARGINAL THEORY

The development of economics in the last years of the nineteenth century is in marked contrast to that of history, in that here the trend was towards deductive theory drawn from abstract first principles, rather than to the accumulation of facts. The 'vulgar economists' of the mid century were content with asserting the iron laws of economics as a scientific justification of the harshness of exploitation of their time. However, already in 1852, in his *Principles of Political Economy*, even John Stuart Mill had begun to doubt whether this degree of misery could be justified by any laws, and to show an alarming tendency towards socialism.[7.9] Indeed in the mid century the basic purpose of economic writing had changed radically though imperceptibly. The task of the economist of the new period was no longer to defend capitalism against old-fashioned protectionists and the landed interest, but against a criticism rising from below, from the socialist movement and, most explicitly, from Marx, and this required a more subtle and scientific justification.

The line of this justification, first used by Jevons (1835–82), Menger (1840–1921), and Walras (1834–1910), and built into a complete system by Marshall (1842–1924), was the theory of *marginal utility* (p. 1061). It was scientific only in the sense that it applied a slightly more refined mathematical theory of limits to the problems of economic exchanges, while accepting the essential and timeless naturalness of the capitalist scheme of production. In this theory all values are fixed by that at which the last or marginal unit can profitably be exchanged. These marginal units: the field that is just worth cultivating; the extra article that it would be just worth manufacturing; the man that would be just worth taking on or sacking; the purchase or the enjoyment one would just be prepared to do without, are taken as the limiting points at which production or consumption of particular commodities starts or ceases. Average costs are replaced by marginal costs, the difference between them being considered to furnish both the explanation and the justification of rent, profit, and interest.

The marginal theory denied any significance to *value* other than that determined by the law of supply and demand in an ideally free market. Value depended solely on a purely subjective mathematical calculus, independent of vulgar material considerations. Accordingly, as labour does not enter into the determination of value, there can be no question of surplus value or exploitation, so that the arguments of Ricardo and Marx are both regarded as irrelevant. The marginal theory also substituted for the pre-existing harmony of the eighteenth-century economists a *perfect equilibrium* which is absolutely automatic in action but one which is at the same time an optimum. Any interference by trade union, monopoly, or government action with exchange at the true marginal values inevitably produces a worsening of the situation for everyone.

The marginal theory was not presented in the first place with this justificatory intention. It was an attempt to account for price changes entirely in the terms of the market and stock exchanges, neglecting any consideration of the productive process itself, in which the economists of the time, with the exception of a few economic historians, were not seriously interested. The theory had never had, in any case, much contact with economic realities. Business men made no use of marginal calculation in their transactions. The actual prices of particular goods were subjected to chance fluctuations due to external causes or speculators, which the theory ignored, and at the same time prices in general were subjected to the much more serious, seemingly inevitable, but unpredictable variations in booms and slumps. These could not be ignored, but were treated as oscillations which would disappear when ideal equilibrium was reached.

Long before the marginal theory had been accepted the conditions it postulated no longer existed. The free world market was visibly being encroached upon by rings and trusts on the one hand and by protectionist governments, often closely allied to them, on the other. The spread of heavy industry into areas outside Britain (p. 562) had, in fact, removed the highly favourable mid-century conditions on which the marginal theory had been based. Yet it remained the economic theory that became established doctrine, not only in Britain but in nearly every capitalist country, and was taught with minor modifications in every department of economics. The very fact that it had little contact with reality only set off the beauties of its scientific reasoning. It was left to Lionel Robbins, making an explicit virtue of its failings, to declare:

... both individual valuations and technical facts are outside the sphere of economic uniformity. ... But is it not desirable to transcend such limitations? Ought we not to wish to be in a position to give numerical values to the scales of valuation, to establish quantitative laws of demand and supply? ... No doubt such knowledge would be useful. But a moment's reflection should make it plain that we are here entering upon a field of investigation *where there is no reason to suppose that uniformities are to be discovered.*

329. The works at Essen of Alfred Krupp (1812–87) about 1880. In 1848 Krupp had obtained control of an iron forge employing three men, founded some 38 years earlier by his father, and died leaving a works in which 20,000 people were employed. This was symptomatic of the spread of heavy industry outside Britain. From a contemporary woodcut.

. . . If this is true of attempts to provide definite quantitative values for such elementary concepts as demand and supply functions, how much more does it apply to attempts to provide 'concrete' laws of the movement of more complex phenomena, price fluctuations, cost dispersions, business cycles, and the like.[7.135]

And to boast that true economics involved no

tedious discussions of the various forms of peasant proprietorship, factory organization, industrial psychology, technical education, etc. . . . [or] spineless platitudes about manures.

Nevertheless, the events of the twentieth century, particularly the great slump of the thirties, proved too much even for academic economists. How the revolt of the most eminent of these, Maynard Keynes (1883–1946), against its defects and absurdities marked the beginning of a new era in economic theory, involving the principle of full employment, will be discussed in its place (pp. 1134 ff.).

The major features of the marginal school, its formation and subjectivity, are characteristic of the general intellectual retreat of the late nineteenth century. It marks a return to the misplaced mathematical interpretations of social phenomena which occurred at the end of the seventeenth century. With the introduction of statistics it appeared that, in Jevons' phrase, 'our science must be mathematical simply because it deals with quantities'. This elimination of the social and the material from economics appealed to the intellectual because it appeared to be *scientific* and *objective*, in the positivist and fundamentally subjective sense which was growing more important towards the end of the century (p. 570). It was, in fact, neither objective nor politically neutral. By putting all emphasis on the consuming individual and his subjective wants, it was showing that there was nothing to choose between the desire of a millionaire for his second Rolls and that of a workman's wife for a pint of milk for the children. By eliminating any concept of exploitation from economics it put revolutionary criticism out of court. By taking the existing economic system for granted it came ultimately to justify it. That justification was certainly badly wanted and was becoming more and more difficult.

'FIN DE SIÈCLE' STERILITY AND DECADENCE

The foretaste of trouble, which began with the great depression of 1870 with its business insecurity, labour unrest, and fears of war, was to have an even more direct effect on the social sciences than it had

on the natural sciences (pp. 664 f.). In spite of ever-mounting accumu-
lations of wealth, indefinite progress did not seem so certain, and even
its direction was in question. Even in Britain at the height of its prosperity, vaguely socialist tendencies grew and spread into the middle
classes. In reaction against this, the general tendency in academic and
intellectual circles was to ignore or to attempt to explain away the
evils and dangers of civilization as merely apparent, and to distract
attention from the social to the personal and psychological sphere. In
an age of admitted decadence there was a trend towards a religious,
mystical, and frankly irrational social science. Philosophers like
Nietzsche and Bergson, and sociologists like Pareto (1848–1923) and
Sorel (1847–1922), pointed to the helplessness of man in all his rational
collective efforts, and to the need either to take frankly violent and
irrational action, or to withdraw to absolute inaction and mystical
contemplation. If the social world was beyond the sphere of reason
its improvement must depend on the inspiration or intuition of the
genius or the superman.

Nevertheless, the profound intellectual advances brought about by
the rise of physical and biological science, and the association of the
prevailing liberalism with a strong tradition of rationalism, made it
impossible for a long time to secure the general acceptance of these
reactionary views. Some scientific justification had to be found so that
the social sciences, psychology in particular, could be refurbished with
a new scientific mysticism instead of the old religious one. Many of the
most advanced, as well as the most respectable, scientists and philosophers contributed to this new mystification, ranging from Poincaré
(1854–1912) in France and Bertrand Russell in England to the hard-
headed William James (1842–1910) in America, the founder of pragmatism and brother of the subtle and involved novelist Henry James.

PRAGMATISM

Pragmatism was the first major contribution of the New World to social
thought since Franklin. Its quality may serve to show how much
culture and the social conscience had decayed there under the influence
of unrestricted capitalism. In the 'Gilded Age' of Mark Twain[7.166] a
fantastic exploitation and waste of natural resources was taking place,
and the almost open warfare between rugged individuals was resulting in the success of a few millionaires who were founding the big
trusts.[7.186] Its social character was shrewdly though somewhat superficially analysed by Veblen (1857–1929).[7.171] It was an atmosphere in

330. High Society in the 'Golden Age' of New York. The Gibson Girls from the *Century Magazine*, 1893. It was in 1896 that William James's *Will to Believe* was published, to be followed eleven years later by the more formal *Pragmatism, a New Name for Some Old Ways of Thinking*.

which cupidity, violence, and religiosity were precariously balanced.[7,8] William James felt that an infusion of science was needed to keep them within bounds. He took his science from evolutionary biology, particularly from the abuse of the doctrine of the survival of the fittest. It was good to survive, therefore what led to survival must be good and the way itself must be the true one. Truth was what worked and paid off. This doctrine had the convenience of throwing the cloak of philosophic approval over church-going and money-making alike. It was, thanks to James's most famous disciple, John Dewey (1859–1952), destined to be the basis of American liberal thought and education. For all its 'advanced' character, it was scientifically empty and morally bankrupt, and it was later quite inadequate to resist the openly reactionary and obscurantist trends of Americanism.

THE AUSTRIAN SCHOOL

The most notable contributions to producing a world outlook which could be intellectually tenable and yet ran no risk of coming into conflict with the forces of capitalism came from Vienna. Until the

coming of the war Vienna was enjoying its last years as the capital of
the hopelessly backward and unstable Austrian Empire, which also then
contained the intellectual centres of Budapest and Prague, all with a
common culture, which, though essentially German, was free from the
aggressive self-satisfaction of Berlin. This culture was the exclusive
possession of a limited circle of disillusioned intellectuals of many
nationalities, mainly dependent directly or indirectly on the imperial
administration. It was accordingly a most eminently suitable city for
the creation of a school of intellectual pessimism on all fronts. Both
in philosophy and psychology it produced contributions, neo-positiv-
ism and psycho-analysis, which were to have a profound effect not only
in the academic field but on the whole orientation of 'Western' or
bourgeois thought in the twentieth century. Nor was the school confined to these fields; Menger had founded an Austrian school of
economists which, in parallel with the British school, developed the
marginal theory with an even greater insistence on the logical impossibility of socialism, which was already becoming a serious force in
Vienna.

MACH AND POSITIVISM

The first major contribution of the Vienna school was the construction
of the philosophy of *positivism* (p. 570), largely due to Mach, which
asserted that science was simply the most convenient mode of arranging sense impressions, and that any discussion of the real material
world was pure and useless metaphysics. The early positivists were
particularly scornful of material concepts such as those of atoms,
unfortunately for them just at a period when physical research was
giving those concepts a reality which has become more and more
apparent with the years.

Though positivism appeared in the first place in the field of physical
science, its implications were far wider. As we have seen in the discussion and the development of physics (p. 746) and of the marginal
theory of economics, it marked a general retreat or withdrawal of
intellectuals from concrete to abstract problems and from a naturalistic
to a formal approach. Underlying that movement was a reluctance to
face facts; for facts, especially social facts, were becoming increasingly
hard for bourgeois intellectuals to face. Positivism did indeed provide an admirable alibi for those who wished to be above the battle
on the right side. As the Irishman said, 'I know you were impartial
in this fight, but which side were you impartial on?' Only the most
stupid reactionaries objected to it: unfortunately for the Austrian

positivists the Nazis were among these. On the other side, positivist ideas easily penetrated the intellectual wing of the Socialist movement, increasing its detachment from unpleasant realities and weakening its will to action. A fuller discussion of the development and effects of positivism, which were more evident in the twentieth century, will be given later (pp. 1160 f.).

FREUD AND PSYCHO-ANALYSIS

The second major contribution of the Vienna school was the apparent revolution in psychology brought about by psycho-analysis, with its exposure of the hollowness and bias of conscious reason and its emphasis on the irrational and amoral unconscious mind. By the end of the nineteenth century it was clear that the abstract faculty psychology of the schools was useless except for teaching. It was essentially Aristotelian, and had deliberately been kept so because all questions of feelings or morals had to be left to the care of organized religion. A new 'scientific' psychology was needed, and it was ultimately provided by Sigmund Freud (1856–1939) in the years after 1890. As, however, many of the developments and most of the influence of Freudian psychology belong to the twentieth century, it will be better to treat it in the next chapter (pp. 1154 f.).

POSITIVISM AND REACTION

The whole tendency, exemplified in the Vienna school and in parallel movements in Britain and America, was to find philosophical, psychological, and economic justifications for capitalism, or, as their followers would now prefer to call it, for 'Western civilization', that would give the appearance of scientific analysis in the liberal tradition. By stressing in the most learned terms the intrinsic irrationality and meaninglessness of the universe, the inevitability of personal bias in our interpretation of it, and the irresponsibility of the individual psyche at the mercy of its complexes, they confused and paralysed attempts to control and transform it. By insisting on the rights of the individual to remain detached from society, they furnished a haven for those who wanted to justify their own ineffectiveness.

SOCIAL THOUGHT IN BRITAIN

In Britain, in accordance with traditions that go even farther back than the Industrial Revolution, abstract theory never played such an important role as in Europe, and, accordingly, social science and practical politics were never entirely distinct. The British, more particularly the English, for Scotland still clung to the European tradition, prided themselves on being practical and having no use for theory or metaphysics, especially on social matters. As we have already seen in discussing the physical sciences (p. 746), this was only an illusion, for much theory, most of it unacceptable if it had been brought into the open, was implied in their practical judgements. Nevertheless it was a convenient fiction up to a point, for it could conceal any degree of internal inconsistency and it allowed for change without loss of face. It was by such methods that, as long as outside circumstances remained favourable, the English bourgeoisie had, for the greater part of 300 years, managed to have their way without the violent collisions which their colleagues abroad had to suffer. From 1850 to 1880 they had indeed to meet very little opposition from the working class.

The craft unions – engineers, woodworkers, etc., which were the first to be formed after the collapse of Chartism (1848) – eschewed political aims, apart from a general support for liberalism, and concentrated on the struggle for better wages and conditions. The organized workers accepted the economic system as it was, only asking for a fair share of the product of their labour. As long as Britain continued to be the workshop of the world that share remained ample enough to prevent them from following the example of the Continental workers, who were coming under the influence of Marxist social democracy or of anarcho-syndicalism. Already, however, in the latter nineteenth century the first signs that circumstances in Britain were not so favourable were becoming discernible. The depression of the seventies had been the signal for the revival of trade union activity. The idea that private enterprise and self-help were bound to produce steady progress and prosperity seemed to be shattered by experience. The unskilled workers, whose conditions were becoming intolerable, began to form unions. With the appearance of organized dock workers and gas workers the general character of trade unionism began to change. Older unions, such as the miners', started to reorganize and put up a fight. Not only were economic demands pressed home in hard-fought strikes, but also the justice of the capitalist system was again called into question.

Socialism began to be studied and propagated by working men and intellectuals. The influence of Marx was to be found first in 1883 in a diluted and opportunist form in the Social Democratic Federation of H. M. Hyndman (1842–1921), largely concerned with parliamentary representation. Later a break-away body, more definitely propagandist

331. Dockers applying for work, 1886. Labour at the docks was on a casual basis and three years later there was a dock strike to obtain a guaranteed wage of at least sixpence an hour, but it failed. In 1886 there were riots of unemployed, and the formation in 1893 of the working-class Independent Labour Party with Keir Hardie as president was not surprising. From the *Illustrated London News*, 1886.

and Marxist in character, was the Socialist League of William Morris (1834–96), who brought into the movement his passionate faith in human capacity for fellowship and the creation of beauty.[7.161] As he expressed it:

> The contrasts of rich and poor are unendurable and ought not to be endured by either rich or poor. Now it seems to me that, feeling this, I am bound to act for the destruction of the system which seems to me mere oppression and obstruction; such a system can only be destroyed, it seems to me, by the united discontent of numbers; isolated acts of a few persons of the middle and upper classes seem to me quite powerless against it; in other words the antagonism of classes, which the system has bred, is the natural and necessary instrument of its destruction.[7.117]

In 1893 a more definitely working-class organization, the Independent Labour Party, was formed with Keir Hardie (1856–1915) as president. It aimed at getting working-class candidates into Parliament on a programme 'to secure the collective ownership of all means of production, distribution, and exchange', one far in advance of what passes for socialism in Labour Party circles today. Nevertheless its policy remained an essentially reformist one rejecting any suggestion of revolution.

SOCIAL SURVEY

To most liberal and forward-looking people the need was not so much for rebuilding society as for removing the worst blemishes that disfigured it. What these evils were could be precisely determined by the development of a new method of investigation: the social survey. Its antecedents were the Royal Commissions, which had all through the nineteenth century thoroughly investigated many aspects of British economic and social life. Usually little or nothing was done as a result, but the reports were a mine of social information, of which Engels and Marx had made full use.[7.114]

The new social surveys represented a similar investigation undertaken privately and carried out by scientific methods, but usually with a political and economic objective in somewhat distant view. This was first and most fully developed in Britain, still the leading capitalist country. Following the example of the early socialists, first Booth (1840–1916) and then Seebohm Rowntree and the Webbs started to study in a systematic and statistical way the actual condition and pattern of life of the poor and the working class. The motive was first of all philanthropic, but almost from the start it developed a delicate political flavour.

THE FABIAN SOCIETY

In 1884 Sidney Webb (1859–1947) brought together a group of civil servants, philanthropists, publicists, and other eminently well-meaning people, including H. G. Wells and G. B. Shaw (1856–1950) and founded the Fabian Society. Their main objective was reform of the administrative and economic system through 'measurement and publicity', and their very title indicated that their objective was to be obtained through 'the inevitability of gradualness'. One of their main concerns was that the transformation of society should take place by persuasion and without revolution, so that socialist measures could be adopted by Liberal or even Conservative administrations. Much was done in exposing the horrible conditions of the slums and sweated industries, much in the improvement of local government machinery, while in their *History of the British Trade Unions,* the Webbs gave an intellectual
D

332. John Mills (Mr Polly) in a scene from the film *The History of Mr Polly*, based on the book by H. G. Wells, whose novels influenced millions.

and moral support to the largely spontaneous organization of the working class. The Fabian tracts,[7.51] some of which contain writing by Bernard Shaw[7.176] worthy of a better cause, spread the ideas of this diluted socialism far wider than the Society itself. Shaw's plays and Wells'[7.174] novels were later to influence millions, and create an atmosphere in the early twentieth century in which Fabianism was almost taken for granted in intellectual circles.

FABIAN SOCIOLOGY: THE LONDON SCHOOL OF ECONOMICS

The characteristics of Fabian sociology were avoidance of fundamental theory and attention to facts, preferably facts in large numbers. Conclusions were drawn only from events that were safely buried in the past, as in the magnificent studies by the Hammonds of the town and country labourers of the eighteenth and early nineteenth centuries.[7.65-6] But the problems of the day were to be dealt with piecemeal and slowly. The force behind the improvement was on no account to be an insurgent working class, but rather an enlightened civil service and business community sufficiently aware of the dangers of revolution to prefer the gentler alternative, which was intended to be the indefinite continuation of a progressively reformed and ever more rational and tidy capitalism.

The Fabians had, and were proud of having, no economic theory of their own. Instead they accepted 'scientific' economics, that is the marginal theory, without seeing or perhaps without wanting to see that capitalism cannot be changed by the use of a theory which implies its indefinite continuance. One of the greatest achievements of the society was its promotion of the foundation of the London School of Economics and Political Science in 1895, the original object of which was to train generations of young men in progressive thought and enlightened ideals. With men of the calibre of Graham Wallas (1858– 1932), Hobhouse (1864–1929), Tawney (1880–1962), and Laski (1893– 1950) on the staff, it might have been expected to produce, if not a social revolution, at least a re-interpretation of economics and sociology. Actually the intrinsic timidity of Fabianism prevented it doing anything of the kind. Instead it produced competent administrators and future Labour ministers. Political science was to be little more than reassertions of edifying views on democracy and liberty.

Fabianism was in essence nineteenth-century liberalism adapted to the conditions of the imperial period. It recognized the inadequacy of the old *laisser-faire* policy in an era of growing monopoly and State interference. Instead of fighting imperialism it set out to make it more intelligent and efficient. It did not see, or refused to look at, the inherent contradictions of imperialism that were to lead to wars and slumps. When the Boer War broke out the society was almost split in half. Pacifists, liberals, and ILP members, including J. R. MacDonald and Mrs Pankhurst, protested against the war but the official view sponsored by Shaw and the Webbs prevailed.

The Fabian movement has been treated at such length not because of any contribution it has made to social science, but because it represents, on the highest intellectual level, an effort to cure capitalism by understanding and kindness or, to put it differently, to run away from the duty of getting rid of it. It is still important because it provides the intellectual base in Britain, and to a certain degree in the United States, for the Labour Party and the Progressive Democrats, and is similar to, but not identical with, the Reformist Social Democrats of the continent. Though Bernstein, the originator of the Marxist revisionist movement, had been in England and had come under the influence of the Fabians, that movement turned back from Marxism only when there was any danger of its coming into conflict with capitalism. The Fabians had avoided even that risk from the start.

THE INNER CONTRADICTION OF SOCIAL SCIENCES UNDER CAPITALISM

Even such a brief story of the development of the social sciences in the latter part of the nineteenth century must reveal their incoherent and contradictory nature. It was in fact more and more difficult to reconcile the drive for greater knowledge and understanding, made necessary by the increasing scope and complexity of the social scheme, with the ban on the dissemination of anything that might upset the ruling order based on the exploitation of labour for profit. Such science must necessarily be partial, compartmented, and apologetic in tone. All the real issues were evaded the more easily by being so copiously hidden in a cloud of disputes about trivialities. Academic social science in a decaying capitalism could have no more future than could the social order that had called it into existence and directed its steps.

12.9 The Growth of Marxism in the Nineteenth and Early Twentieth Centuries

During the same period, from the creative work of Marx in the forties to the success of the Russian revolution in 1917, a radically new and quite unofficial social science had been growing up and flourishing. The Marxists, persecuted, without money, without any of the apparatus of Press, literature, science, or education that were available for propagating official views, kept up all through the nineteenth and early twentieth centuries a continuously developing criticism of society, one that was not only far deeper but also far wider than that of academic social science.

One of the key ideas of Marx was the unity of the sciences, and, accordingly, Marxism developed without that separation between science and the humanities or between natural and social sciences which became more and more characteristic of bourgeois culture. Economics, technics, and the physical sciences were all combined with the social sciences in one active unity, which found expression in Marx's *Capital*, of which the central economic doctrine of capitalist

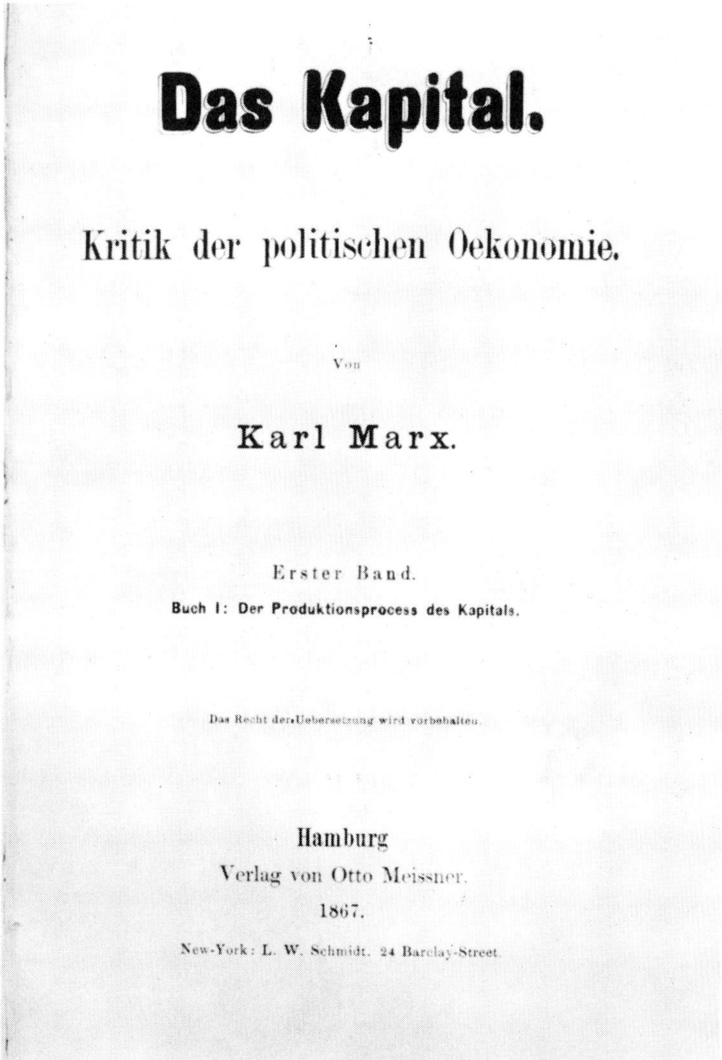

Das Kapital.

Kritik der politischen Oekonomie.

Von

Karl Marx.

Erster Band.

Buch I: Der Produktionsprocess des Kapitals.

Das Recht der Uebersetzung wird vorbehalten.

Hamburg

Verlag von Otto Meissner.

1867.

New-York: L. W. Schmidt. 24 Barclay-Street.

333. Karl Marx was born in Prussia and studied jurisprudence, philosophy and history at Bonn and Berlin. He later studied sociology and political economy and in 1843 he moved to London after moves to Cologne, Paris and Brussels. His *Das Kapital*, the title page of which is shown here, was written in the Reading Room of the British Museum.

exploitation through surplus value has already been set out (pp. 1073 ff.). This is, however, only a part of the wealth of economic knowledge it contains. In addition, not only did Marx carry through a deep analysis of the current transactions of industrial production and finance, but he also explained the recurrent crises of over-production to which capitalism was liable. It was more than fifty years before these crises were officially admitted by bourgeois economists to be necessary consequences of the economic system of capitalism and to this day they can neither account for crises nor can they prevent them in one form or another (p. 1135).

DIALECTICAL MATERIALISM

But *Capital* is not only an illuminating and damning account of a social system. It contains, in its general construction and on almost every page, examples of a new method, as important in the development of historic and social sciences as was the experimental method of natural science of the seventeenth century. Both have a common base in the world of matter and in man's experience of understanding and controlling it; but the method of Marx, *dialectical materialism*, is a wider generalization applicable to the whole field of human endeavour, the world of man as much as the world of Nature. Armed with this method, Marx and his followers attacked numbers of problems that had proved insoluble in terms of the old philosophies, and many new ones to the very existence of which they had been blind.

It is impossible within the scope of such a history as this to give any account of dialectical materialism adequate for those not familiar with it;[7.11.365] what follows is hardly more than a catalogue of its main features. Its essence is the understanding of the appearance of *new forms* and *new processes* as a result of internal conflicts, rather than as an effect of external forces alone (mechanism) or of the influence of a predetermined end (teleology). Marx had found his prototypes of dialectical materialism in the social field, in transformations such as that from feudalism to capitalism and in the much more important transformation in which he was to play such a large part: that from capitalism to socialism. Here internal contradictions and the appearance of new forms were far easier to observe than were changes in the inanimate world or in organic evolution, with their far slower rate of change.

Marx made use of the abstract Hegelian form of the transformation from the categories of *qualitative* to *quantitative* to describe real changes in a complex material system. He showed how a new system of society, qualitatively different from the old, had arisen naturally as the effect of an accumulation of quantitative changes in the old: such as the growth of trade or of machinery with their associated social conflicts. Moreover, the qualitative change had to be abrupt and decisive, for unbroken development could never lead to anything really new. The dialectic accounts for changes by showing how they arise from the stress of conflicts generated inside the earlier system and ultimately breaking it down to replace it by a new system. This approach does not need to invoke either determining material factors such as climate or disease, operating from outside, or to fall back on the concealed ignorance of attributing all changes to providence or chance. A discussion of how this applies to the development of societies through the interplay of productive forces and productive relations has already been given (pp. 1072 f.), but it needs some further explanation here.

According to the Marxist picture of social development, with each change of the economic system was associated the appearance of a new class in society, with very different shares in the work and in the appropriation of its products. Each new dominant class would in the process of its exploitation of other classes appropriate the largest share of the product. In doing this they tended to leave the actual processes of production more and more to other classes. The class that came to operate the most advanced methods of production of the time would, after a struggle, supplant the older dominant class. In doing so it would change the productive relations of society to conform with the new methods of production, with a corresponding change in ideology. Thus the ruling plutocracy of the classical slave States gave way to a largely barbarian feudal nobility more adapted to the running of a subsistence economy (p. 253). The feudal lords yielded in turn to the bourgeoisie, who had fought for and won the right to trade and manufacture in an exchange economy (p. 411), while now the bourgeoisie is being displaced by the workers, whose socially co-operative labour belongs to the era of scientific production and planning (pp. 717 f.). It was in this sense that Marx claimed that all history was one of class struggle.

Thus the dialectic of history was not a formal pattern imposed by Marx on the development of society, but arose from the actual analysis of its changes, and only then was it made the basis of wider generalizations. What makes Marx one of the greatest scientists was his realization that the prime task of a student of society was to discover the inner laws of its development and that having found them himself, he went on to prove their efficiency in practice as well as theory.

ENGELS: SOCIALISM, SOCIETY, AND NATURE

Engels expounded Marx's work and extended its implications outside the field of economics. In *Anti-Dühring*[7.45] he sketched the pattern of a scientific socialism, not derived from wishful thinking about an ideal community, but as a natural dialectical development of the working class in destroying the bourgeois capitalist State and building their own socialist State in its place. In *The Origin of the Family, Private Property, and the State*[7.49] he extended the same method to the early stages of human history. He saw the importance of Morgan and Tylor's studies of primitive societies (p. 1081), and recognized in the economic factors of private property and the division of classes the forces that converted a community of equals, living by unwritten tradition, into a civilized State with its rich and poor, its laws, police, religions, and wars. Finally, in his *Dialectics of Nature*, unpublished until our own day,[7.46] he reached out to explanations of the origins and transformation of pre-human nature, showing how the same dialectic principles also held sway there. It is only now that we are beginning to understand how

334. Friedrich Engels (1820–96) was born in Bremen. From 1842 onwards he lived mostly in London and Manchester and worked in London with Karl Marx.

powerful a method Marx and Engels had forged, and how badly it is needed to make the world of Nature and of humanity intelligible as a process.

CHANGING THE WORLD: THE FIRST INTERNATIONAL

If Marx and Engels had been content merely to put forward their theses on economics, history, and philosophy as academic exercises, they might have won a place in the learned world in spite of their radical conclusions; but from the very nature of their approach to society they could not accept these limitations. Marx was not a setter-up of philosophical systems; he did not want merely to contemplate the truths of science, but to use them to transform society. He realized that the laws of social development which he had discovered gave, for the first time, the opportunity for effective and consciously directed social action. What the natural scientists had done in transforming Nature through a knowledge of natural law and the use of natural forces could now be achieved in the social sphere. Changes in society had indeed always been brought about through the action of conscious human wills. There was all the difference between action based on understanding of a total situation and that determined only by the choice of limited and ignorant men, for here the results had rarely corresponded to the original intentions.[7.47] Indeed the major transformations of society had come about without anyone consciously desiring them; so much so that fate, providence, or natural law had to be invoked to explain them.

Thus, while the academic sociologists and economists discussed and explained the workings of the capitalist system, which they considered as the only true, natural, and free economic system, the Marxists were already taking action, welding together the largely unawakened and unorganized working class, and rousing them to a realization of the role they were to play in the transformation of society and in the organization of a new planned economy to take the place of that of capitalism. The foundation of the First International and of the original Social Democratic Parties was not only an integral part of the life work of Marx and Engels; it was an integral part of their development of social theory. The early Marxists created social science while making social history.

Marx and Engels and their increasing band of fellow fighters remained continually in touch with and alive to political and social changes in every part of the world. Every event of importance from America to India, from Japan to Ireland, was subject to their scrutiny and judgement.[7.110] They learned lessons from the successes and also from the failures of socialist and progressive movements. The most famous of these, the Paris Commune of 1871, was started by idealist and liberal patriots. Marxist communists played no leading parts in it. However, from the very moment of the outbreak Marx saw that the solid working-class support given to the Commune and the senseless brutality of its repression by the agents of French capitalism had transformed an unsuccessful assertion of democratic rights into a prototype of the working-class seizure of power. He wanted the working class throughout the world to emulate the courage and avoid the weakness and confusion of the Communards by a more thorough grasp of revolutionary theory.[7.107]

DIFFICULTIES AND DIVISIONS IN THE CAMP OF SOCIALISM

After the death of Marx in 1883 the socialist movement, though growing in strength, lost something of its understanding of events, and there was no marked advance in theory. This was largely due to the emergence of a new phase of capitalism – that of imperialism – which Marx had only begun to deal with, and which was only adequately analysed by Lenin twenty years later (p. 1167). Meanwhile, Marxist socialists had not only to fight against capitalists and against the governments they controlled, they had also to deal with difficulties within the movement itself. The very success of Marxist doctrine among active working-class movements in Europe brought new difficulties in its train. What Marx had given to the working class and their allies was the reasoned belief in the possibility of the overthrow of capitalism. He had done something as well to sketch the kind of socialist system that would take its place.[7.109] At the same time, however, following the Marxist programme required an exercise of understanding, effort, restraint, and discipline too great for many who were in agreement with the general objectives of socialism. They wanted short cuts or easy ways to socialism. There were splits and rival movements, and much of the time and work of the early Marxists was taken up with apparently endless but essential controversy which helped to extend and deepen the understanding of society.

In the first place there were rather shallow and well-meaning persons with a genuine feeling for social betterment who attached themselves to the band-wagon of socialism, but who completely lacked the knowledge or political sense needed to form and direct an effective socialist movement. In particular, they avoided the whole idea of class struggle and imagined that all that was necessary to achieve socialism was to

335. The Paris Commune of 1871 was brutally repressed and prisoners were shot. From the *Graphic*, 1871.

set out a sufficiently attractive picture of what it was like. Of these in Britain there were the Christian Socialists and the followers of Henry George (1839–97), and in Germany the 'true' Socialists like Herr Eugen Dühring, who has been so completely dealt with by Engels that he may stand for the whole tribe.[7.45]

ANARCHISM

More serious were the defections of those who, having once worked alongside the Marxist socialists in a common struggle against capitalism, had broken from them and taken different paths. The first of these defections, that of the anarchists, was the result of the break away of the impatient and violent Russian *émigré* nobleman, Bakunin (1824–1901), which broke up the First Working Man's International in 1876. Anarchism had a special appeal to generous sentiment because of its denial of the need for tedious social theory, either before or after the destruction of hated governments; and because of its violent anti-clericalism, its contempt of legality, and its romantic revolutionary flavour. It found its leaders among the *émigrés*, and its support from workers in backward countries in southern Europe and in America, where the impact of capitalism had been peculiarly brutal, where there was little large-scale industry which could provide a core of disciplined workers, and where legal and peaceful remedies were not to be had. For all the devotion and sacrifice it commanded, anarchism did not and could not even mitigate these conditions. The best it could do was to maintain the spirit of class militancy; the worst was to furnish a basis for diversions and internecine quarrels, thus weakening the whole movement.

REFORMIST SOCIAL DEMOCRACY

In the countries of more developed capitalism, particularly in Germany, Scandinavia, and Britain, the temptation was not to intensify but rather to escape from the rigours of the class struggle by pretending that it did not exist. Relying on the additional profits that imperial exploitation was bringing in, it was tempting to believe that socialism, even Marxist socialism, could be achieved by an evolutionary process. Nothing more drastic was required than a gradual and imperceptible modification of capitalism through the operation of normal democratic processes. This view completely underestimated the power and tenacity of the capitalists and their willingness to use extra-legal methods of repression, as the events in Germany of 1918 and 1933 were to bear tragic witness. In Germany, where the movement had the most success and where strong Marxist Social Democratic movements had been built up as early as the 1870s, the tendency to revise Marxism, to prove scientifically that it had served its purpose and was now out of date, was most strongly marked. Marx himself pointed out and condemned this tendency in his *Critique of the Gotha Programme* of 1875. His analysis of reformism expressed in the circular letter which he issued with Engels in 1879 is still relevant:

It is the representatives of the petty bourgeoisie who are here presenting themselves, full of anxiety that the proletariat under the pressure of its revolutionary position, may 'go too far'. Instead of decided political opposition, general compromise; instead of the struggle against the government and the bourgeoisie, an attempt to win and to persuade; instead of defiant resistance to ill treatment from above, a humble submission and a confession that the punishment was deserved. Historically necessary conflicts are all reinterpreted as misunderstandings, and all discussion ends with the assurance that after all we are all agreed on the main point.[7.109]

While the warning as to their fate:

These are the same people who under the pretence of indefatigable activity not only do nothing themselves but also try to prevent anything happening at all except – chatter . . . the same people who see reaction and are then quite astonished to find themselves at last in a blind alley where neither resistance nor flight is possible

was to be unheeded until Hitler's time. Marx and Engels made their own position quite clear:

As for ourselves, in view of our whole past there is only one path open to us. For almost forty years we have stressed the class struggle as the immediate driving force of history and in particular the class struggle between the bourgeoisie and the proletariat as the great lever of the modern social revolution; it is therefore impossible for us to co-operate with people who wish to expunge this class struggle from the movement. When the International was formed we expressly formulated the battle-cry: the emancipation of the working classes must be conquered by the working classes themselves. We cannot therefore co-operate with people who say that the workers are too uneducated to emancipate themselves and must first be freed from above by philanthropic bourgeois and petty bourgeois.

Though Marx and Engels succeeded in preserving a solid core of convinced and determined socialists, the reformist trend was continually being reinforced by those who found in the socialist movement a sure path to positions of power and affluence. Indeed after Bismarck had shown the way with Lassalle (1825–64) capitalist governments themselves learned to patronize a socialism that confined itself to words and could be counted on never to do anything serious to interfere with their interests. As soon as one socialist leader went too far and was absorbed into the ranks of the bourgeois parties, as were Millerand in France and Mussolini in Italy, others equally inclined to compromise were always found to take his place.

THE BRITISH LABOUR MOVEMENT

In Britain a similar drift towards respectability occurred; but whereas on the Continent Social Democratic parties were nominally Marxist, if really reformist, most of the organizations which made up the British Labour movement were openly reformist to start with. Though all drew their fundamental ideas on socialism from Marx, Marxism as a theory was repudiated by all but the small Social Democratic Federation. This was in line with the traditional resistance to all theory which the British had always found such an easy way to avoid fundamental criticism of the class struggle. It also derived from the historic

336. The Chartist revolt occurred in the hungry forties. In Stockport in August 1842, during the great 'turn out' of working people, there was an attack on the Workhouse. The economic position of the working class in Britain improved in the 1850s with the country's commercial triumphs. From the *Illustrated London News*, 1842.

fact that Marx's first attack on the capitalist system came just after, and partly in consequence of, the first reaction of the working classes, which was particularly violent and explicit in the hungry forties in the Chartist movement. After 1850 the commercial triumph of Britain as the workshop of the world eased the economic position of the working class. That situation came to an end by 1880, but within a decade the position had largely been recovered as a result of the rise of imperialism, and the interval of economic stress was too short for the influence of socialism to make itself felt. The British working class still tended to identify itself with its own bourgeoisie in a way impossible for the working class in any other country. We have seen (pp. 1098 ff.) how the role of theorists of the socialist movement in Britain was taken over not by the Marxists but by Fabians. In 1912 and in 1926 great strikes gave warnings that the fundamental conflict with capitalism had lost none of its force. Nevertheless, it is not until the mid twentieth century that the loss of Britain's favoured position in the world, after two

337. The 1851 Exhibition was a shop-window display for Britain's new industrial prosperity and was housed in an iron building which was itself an example of new ideas and methods of construction not previously employed on such a scale.

crippling wars, is forcing the working class to realize the danger of reformist socialism.

One characteristic of reformism is the abandonment of any attempt to produce a theory of socialism different from that of academic social science. The idea that science can possibly be influenced by the upbringing, social milieu, or interests of the scientists is indignantly repudiated. The doctrines of automatic and imperceptible biological *evolution* are supposed to have made *revolution* unscientific and unnecessary, while the acceptance of marginal economics seems to remove the very existence of the exploitation on which Marx had relied to provide a motive to abolish capitalism.

The reformist tendency, though it contributed to the retardation of the achievement of socialism in the advanced capitalist countries, was itself a consequence of the movement towards imperialism (pp. 705 f.) on which the powers of monopoly capital had embarked to overcome the difficulties of finding markets. In the process the pressure on the skilled workers who produced the capital goods and armaments was slightly relieved. At the same time, as we have seen, in Britain, the major imperialist power, Fabian ideologists looked along the path of empire for the solution of economic and social difficulties.[7.145]

IMPERIALISM AND WORLD WAR

The nemesis which followed these aspirations had not long to wait. By 1914 the whole of the world came to be carved up between a few great powers, all civilized and relatively prosperous. The world economy, in which the vast majority of mankind were doomed to be producers of cheap raw materials that were then worked up in the advanced industrial countries, was economically sound but it was to prove politically and humanly unstable. Labour troubles increasingly racked the advanced countries while methods of diplomacy found it more and more difficult to cope with the struggles between the imperial powers themselves, which finally exploded in the war. It was in that war that the first break-down of the capitalist system occurred in its weakest point, in Russia. Though temporarily reconstituted after the First and Second World Wars, the old world was in fact gone for good. What this meant to the sciences of society will be discussed in the next chapter (pp. 1115 f).

Chapter 13

The Social Sciences after the First World War

13.0 Introduction

The war of 1914 shattered the hope of a permanent, peaceful evolution of capitalism. While its outbreak marked the end of one period, the 1917 Revolution in Russia marked the beginning of a new era. Socialism, from being a doctrine and a rallying point of resistance to economic injustice, was now being put to the test of actual practice. One of the world Powers had now become a socialist State based on Marxist principles. From then on Marxism, from being an unofficial and outlaw philosophy, was to be the creative basis of thought of 150 million people and was to become, by our time, that of some 1,100 million, or more than a third of the world's population.

From 1917 onwards two functioning systems of society have existed side by side, corresponding to the two systems of thought: capitalist and socialist. In the capitalist world social science has had to adjust itself to conditions of alternating instability and violence. Over the same period in the socialist world the enormous tasks, first of struggling for mere existence, and then of building a material base for a new kind of social life, have presented a continuous series of new problems to social science.

In this chapter we will first discuss (13.0) the political and economic events that influenced the development of social thought in the twentieth century. This will cover the same ground already treated in the introduction to Part 6, but regarded from a different point of view. It will be considered not so much in relation to technical and economic developments as to events directly affecting people's attitudes and modes of thought. This leads to a general discussion (13.1) of the character of the social thought of the period, both in the capitalist and socialist sectors of the world. There follows (13.2–13.5) a detailed analysis of aspects of capitalist social science. Section 13.2 covers economics; 13.3 the application of social science, statistics, social surveys and market and productivity research; 13.4 deals with education, and 13.5 with the more purely ideological aspects of psychology, philosophy, and theology.

From here we pass to a corresponding, though briefer, survey of developments in the socialist countries; section 13.6 on the practice and theory in the social sciences in the Soviet Union and later in the Eastern European socialist countries, and deals with the critical changes that have taken place in China; 13.7 with those of the liberated countries, formerly of colonial or semi-colonial status, in Asia, Africa, and Latin America, making up the now recognized under-developed parts of the world and whose development is liable to recast most of the social sciences. Finally, in section 13.8 something will be said of the possible future developments of the social sciences.

THE SOCIAL TRANSFORMATIONS OF THE TWENTIETH CENTURY

At no previous period in history has mankind gone through such profound changes in such a short time as in the years after 1914. The economic fluctuations, the bitter internal political struggles, the wars and the preparation for new wars, followed each other without respite, and all well within the span of a single lifetime. The thoughts and attitudes of men and women were inevitably affected by this changing situation, even if with some this meant a deliberate turning away from the whole political, social scene.

The capitalist world throughout this period of unregulated growth and disintegration lacked any central purpose or expectation outside the pursuit of profit or mere survival. It is not surprising therefore that in its ambit there should be a great confusion of counsel, which finds its intellectual expression in a variety of theories in social science. In psychology, in economics, in philosophy there have been a host of rival and somewhat ephemeral schools, few carrying any sense of hope, purpose, or even intellectual conviction. Only those who were willing to examine and criticize the whole system under which they lived, and could follow the tradition of resistance to it, could find in their understanding a basis for hope.

In the socialist world the urgent problems of mere existence at first, later of construction and cultural creation, imposed a need for a strong unity of purpose and belief capable of inspiring whole peoples to carry out almost superhuman tasks. To achieve and maintain that unity was not easy. Living as they did for decades under the conditions of war or the threat of war, the peoples of the Soviet Union had to forge their social theories, profiting by the experience of their mistakes and successes, on the basis of the heritage of Marx and Engels.

We have already discussed earlier (pp. 703 f.) the nature and origin of the economic and political changes of the twentieth century. Here it is only necessary to recall a series of events which are only too familiar to most of my readers, and to say something of their effects on determining the atmosphere of social thought.

The general pattern of change in capitalist countries during the twentieth century is dominated by the First and Second World Wars and by the great slump that marked the period between them. In each decade we can distinguish years of social progress and emancipation; in 1912 with the great strikes and the revival of socialism, in 1920 with the post-war reconstruction, in 1936 with the triumph of the *Front Populaire*, in 1946 with the era of liberation and 'one-world' co-operation, in 1953 with the precarious first thaw in the Cold War, in 1960 with the great movements of liberation in Africa and with the first crack in the neo-colonialist domination of the Americas in Cuba. Between them came reaction and disaster.

After each major disaster, in which the rottenness and hollowness of the old forms of society were revealed, there came a movement of hopeful popular resurgence carrying with it a new rush of progressive ideas. These movements, failing for the most part to break through the multiple and deep defences of the capitalist system, died away in confusion and frustration, and gave place to periods of reaction and repression. But, as no degree of reaction can stabilize a system that carries its own seeds of destruction, these phases in turn passed into fresh disaster. Such an abbreviated and schematic account is necessarily an over-simplification, but it is needed to understand the development of social thought over the period.

It would be wrong to see in these alternations any cyclical tendency of history to repeat itself indefinitely. Each phase had its own characteristics, and in succession they mark the persistence of an overriding tendency, the shrinking of the power of capitalism and its inability to find any stable basis.

The effect of the great slump was necessarily different in character from that of the two wars. Without an open struggle, except in Spain, the progressive movement was divided and suppressed. The Nazi movement was itself an exploitation by the big industrialists of popular revulsion against the régime of unemployment and misery which they had themselves brought about. It was indeed by putting illegality in the service of reaction that the popular forces were held down in the years before the Second World War.

That war showed in turn that the upsurge of popular feeling in the thirties had not been crushed, but only held in check, as it burst out with increasing determination and capacity in the era of liberation. For the first time men and women in all countries in the world, inside as well as outside the area occupied by the Nazis and their allies, found themselves engaged in an enterprise with a common object. During the War, and for a short time after it, there seemed to be some hope of building a peaceful and co-operative world in which countries with capitalist and with communist governments, and even those with mixed governments such as France and Italy, could co-operate. The privileged classes in America and Europe soon destroyed this hope. Making use of every interest and prejudice in their own countries and every mistake in the camp of socialism, they saw to it that friendship was replaced by suspicion and hatred. They instigated the Cold War, with its tragic consequences, in Greece, Malaya, Indo-China, and Korea. However, it proved impossible to crush socialism by any demonstration of strength short of an atomic war, which they were not all prepared to face. Meanwhile the strains of the Cold War in capitalist countries, the up-surge of the peoples of the former colonial countries, and the universal demand for peace and disarmament have, if only partially, broken down the division of the world into two parts. Though the Cold War still continues, there are some signs that we have passed its most violent phase. But there can be no safety until real negotiations start on dis-armament and the last vestiges of colonialism are swept away.

INFLUENCE OF EVENTS ON SOCIAL SCIENCE
IN THE CAPITALIST WORLD

The struggle in the economic, political, and military fields could not fail to affect social science, however academic. Nevertheless, because of the class nature of the struggles themselves and of the social background and upbringing of the intellectuals in capitalist countries, this influence had more often than not a strong bias to the Right. From the moment that Marxism had triumphed in one country and had thus become an example as well as a danger, all the influence of reaction was turned against it. Though the intellectuals of the early century had been formed on the tradition of liberalism, they found it hard to resist the combined attack of politicians, Press, and patronage. They were torn by divided loyalties, between ideas and practice, between champion-ing the downtrodden and being good citizens. This, rather than any original sin, is the real source of the *angst* and guilt that afflict those intellectuals who cannot preserve their faith in the people.

In periods of liberation nearly all intellectuals were carried a little way forward on a wave of popular feeling, as in the period of recon-struction after the First World War or of the New Deal in America.

Once the popular forces got into difficulties, however, and reaction took control, most of them tended to become disillusioned and pessimistic. In their efforts to avoid the imputation of being fellow travellers, some were driven increasingly to distorting the idea of progress and even to doubting its possibility. Many were forced back on to the comforting certainties of the past and, unable to change the real world or to prevent it changing in ways they did not like, preferred to withdraw to private worlds of their own. This intellectual reaction, automatic or discreetly sponsored in Britain, has been enforced by every device of informers, the law, and smear campaigns in the United States.

THE BREAK-UP OF LIBERALISM

In these conditions it is not surprising that the general characteristic of official thought under capitalism in the twentieth century is its timidity, far more marked than even that of the last quarter of the nineteenth. It had served well enough to discourage faith and action, but it had lost all capacity to inspire. It is not that inspiration is disparaged. Sermons and speeches continually assert the need for an inspiration for the free (capitalist) world to balance that of communists, who, however misguided, seem to have some faith to live by. So far, however, that inspiration is still to seek. The tone of officially sanctioned opinion remains fundamentally apologetic and defensive.

Here we have the apparent paradox of a period, essentially one of popular development and liberation, in which theories of society are static or retrograde. The old liberal world-picture of the nineteenth century was destroyed in the twentieth by the very experience of its futility in action, though it survived longer in some countries than in others. In its place, and varying according to the situations in different countries, there grew up a variety of other attitudes, ranging from those of social democracy to those of Fascism. Some of the developments that followed the upheaval of the First World War and the Russian Revolution are discussed in the succeeding paragraphs.

THE AMERICAN SCENE

In the great 'have' country, the United States, a 'return to normalcy' was the ruling idea after the upsetting experience of a somewhat remote war of 1917–18. A belief in competition and in the values of rugged individualism successfully concealed the unchecked growth of big business corporations; at least it did so until the slump of 1929–32 called for 'a new deal'. The evident failure of capitalism to maintain its boasted prosperity, which could be put down to no outside cause, and the need for immediate measures to deal with vast unemployment, favoured an upsurge of progressive thought and some progressive action, such as industrial unionism and the TVA (p. 964). But New Deal thought was too concerned with short-term remedies to search for underlying causes in the economy. The liberal capitalist ascendancy was retained, and only mitigated by a provision for rescuing the worst-hit victims of business instability. The New Deal advocates, including their great leader, Franklin Delano Roosevelt (1882–1945), evaded any serious analysis of the economic situation, which would have meant a criticism of its basis in capitalism. They failed to put forward any fundamental programme which could have rallied the people behind them to maintain and enlarge the liberal movement they had started. They were thus in no position to fight against the return of full-blooded capitalist ideas and practice once the state of the market improved. With the exception of a short and partial interlude in the Second World War, the United States fell deeper and deeper into the hands of reaction and obscurantism, with many Fascist characteristics, but still much better covered by hypocrisy than Hitler had ever managed to do with the Nazi movement.

These tendencies, which sustained the cruel and indecisive Korean war and had reflections in the era of the McCarthy persecutions, flared up again in the reaction to Cuban liberation which nearly brought about a Third World War. There are, however, signs of a reaction to them in some relaxation of Cold War international policy and in the measure of popular support that has been given to the efforts of the coloured people of the United States to secure the advantages which Lincoln's abolition of slavery should have brought about a hundred years ago.

REACTION AND FASCISM IN THE 'HAVE NOT' COUNTRIES

Such movements in the 'have not' countries had started earlier, because there, in Italy after the slump of 1921, in Germany after the great slump, the attitude of the people made it impossible for capitalism to remain in power by democratic means. The only alternative, snatched at eagerly by the big capitalists, was the seizure of power by their reliable supporters. This was effected – through officially connived-at violence in Italy or with constitutional forms in Germany – by gangs of malcontents exploiting mass misery and the frustrations of the lower middle class, financed by the directors of the big trusts. It is an important development only negatively, because it marks an open abandonment of the liberal forms with which the more prosperous capitalism of the past had covered its rule. The ideology of Fascism and Nazism was little more than a compound of race hatred and glorification of war, playing on every ancient prejudice and making the minimum of demands on thought.[7.16] Though the actual power of Nazism and Fascism was destroyed, except in Spain, by the last war, Fascist ideas have retained a strong hold and have spread far wider than the confines of Germany and Italy. In one form or another they still remain the last resort of the supporters of capitalism, the one way that they can hope to hold off, if only for a time, the challenge of socialism.

SOCIAL DEMOCRACY AND LABOUR

Only in the semi-successful countries, particularly in Britain, still the greatest imperialist power, and to a lesser degree in Belgium and Holland and the Scandinavian countries, was it possible to cushion the shock of economic disaster, at least for the middle class and skilled workers, by the use of profits drawn directly or indirectly from colonial exploitation, and thus to allow the development of liberalism into a variety of socialism considerably milder than that which the Fabians had proposed (pp. 1097 f.). The force behind that development was the pressure of working-class militancy. This was already evident in the unrest of the years immediately before the First World War, and had later been encouraged by the example of the Russian Revolution. It took the combined efforts of out-and-out reaction and the restraining hand of social democratic leaders to keep it down in the troubled days after the First World War.

In Britain, where the working class was most strongly organized in trade unions, the drive to achieve a society with fair shares for all lasted longer than elsewhere. It was checked but not stopped by the betrayal of the General Strike in 1926, and has survived the failures of four Labour Governments. Nevertheless, the experience of the last forty years has in Britain, and to a far greater extent in France, which in this respect was more in the position of a 'have not' country, led to a gradual loss of faith in merely liberal progress, that is, in any social progress compatible with the retention of a profit-making economy. It seemed indeed to many middle-class people that there was no longer any better future to look forward to. The most that they could hope for was the preservation for some time longer of just tolerable conditions. The impulse to change tended to come primarily from left-wing Labour circles.

COLONIAL ASPIRATION AND REVOLT

There remains to be considered the development of social ideas among the three-quarters of the world's population – the hundreds of millions in Asia, Africa, and Latin America – who lived through most of this period under colonial or semi-colonial régimes. For the first time in modern history these peoples were beginning to make an important contribution to world culture and politics. Everywhere they were in revolt or near it. The revolts were at first easily suppressed, but later in one country after another – India, Burma, Indonesia, Indo-China, Egypt, North Africa, Central Africa, Cuba – some measure of national independence was achieved. It was natural that the liberal and nationalist ideas of the nineteenth century should at the outset prevail among the colonialized peoples and those of China, which had been for a century at the mercy of foreign imperial powers. The movement was at first limited to the middle classes. But as time went on, and especially during and after the Second World War, the whole people, peasants and workers alike, became more involved in it, and demands for economic planning and internal social justice became more insistent.

In these newly-liberated countries, rather than in the disillusioned and decadent capitalist world, there was a lively hope of better things, if only because for most of the people nothing could be worse than the present. Here also the example of the Soviet peoples, especially those of the Asian republics, and more recently of China, has been most keenly felt. For many of these, from a material level equal to or lower than that in colonial countries, have built up by their own efforts an advanced mechanical civilization without losing their national culture.

On the other side, and particularly after the Second World War, colonial repression grew in ferocity as its scope was diminished and its right more confidently and effectively challenged. The savage, prolonged and unsuccessful wars in Indo-China and Algeria and the increasing repression in Portuguese Africa and in the South African republic have shown that the forces of national liberation cannot be effectively kept down in twentieth-century conditions. That lesson has been, up to a point, learnt by the old colonial powers, but not yet by the new colonial power which has grown up beside them, the United States of America, as witness its reaction to Cuba and Vietnam.

The pressure of national movements has resulted in a great wave of liberation of colonies, particularly those in Africa. The new governments of those colonies that came from the old direct imperial rule increased the number of independent states in the United Nations from sixty to over a hundred, but the degree of independence is very varied. In some ex-colonies, such as Malaysia, bases and foreign troops are still retained; in others, such as Ghana, almost complete economic as well as political independence has been achieved. Over all of them, however, hangs the shadow of the old colonialism, in what is now recognized as neo-colonialism – the retaining by one or more capitalist countries of complete economic, and a measure of cultural, control through owner-ship of mines and plantations and manipulation of the prices of the local products. Such control has long been exercised, and still is, over most of the republics of Latin America with the exception of Cuba, which proves the rule. Nevertheless, over all these vast territories, in-cluding some two-thirds of the human race, living in under-nourished and intolerable conditions, there is an enormous ferment. Peoples are no longer willing to accept from fate a condition so far inferior to what they observe as that of the most developed nations. They are demanding that the gap between their standards and the others shall be diminished, as they know scientifically it can be. All this is bound to have reper-cussions on the social sciences, not only in those countries but also in the industrial countries themselves. This corrective is badly needed, for they have been imbued for many centuries with the mentality of slave-owning master races.

13.1 General Character of Social Thought in the Twentieth Century

Despite the varieties of influence in the different countries and during the different phases of economic political developments since 1917, one common element of struggle runs through the whole period. It has witnessed a long running battle between the insurgent lower classes and suppressed nationalities on one side and the ruling class of the large industrial capitalist countries on the other in which, despite many set-backs, the new forces seem to be gaining the upper hand. Everywhere the success of the people themselves in ending the rule of the rich, first in Russia, then in China and other countries, has remained as an un-deniable historic fact, raising hopes and giving greater confidence to the peoples of all the rest of the world.

If, however, we turn from popular feeling to the expressed views of intellectuals, and particularly of social scientists in the capitalist world,

338. A Red Army meeting at the south battle front where the forces headed by General Denikin were defeated. Some of the results of this victory may be seen, for example, in the illustrations on pages 863, 870 and 874.

the picture is not so simple. They have lived throughout the whole of the period, and especially through the oppressive Cold War phase, under considerable pressure, direct and indirect, to reject the whole picture of popular liberation, to resist all moves that might lead to communism, and to rally to the defence of capitalism under the more attractive guise of the 'free world', 'democracy', or 'Western Christian civilization'. They are not under any obligation to believe in these things, but almost everything in their environment – books, films, broadcasts – urges them in this direction, and it is far more comfortable for them spiritually and materially if they can bring themselves to accept the prevailing values.

Every intellectual naturally wants to be able to carry on with his chosen work, to be able to publish it freely, and to receive some credit for what he has done. Finally, or first of all, he needs to live. A steady safe job, radio contracts, good royalties or film rights, pleasant trips to foreign countries, are, if not absolute necessities, the essentials of a good life. If one has naturally from upbringing and education the opinions that make all these things easy to get, so much the better, but even if not it is most tempting to assume them. With all the resources and rewards of publicity in the hands of government or big business, to stand out against them is frustrating to the last degree, even if there are not yet in all countries positive sanctions in the form of smears and committee hearings which in the United States effectively ruin the career of anyone wandering from the path of loyal Americanism.

The knowledge that these are the alternatives presents an agonizing choice which few are prepared to face; for knowingly to choose ease and honour at the price of integrity would be unbearable and would poison all enjoyment. It is far better to evade the choice altogether and convince oneself, which is not very difficult, that the easy path is also the right one. To take it with a good conscience it is necessary only to use everything that can be found – and there is plenty offered – to discredit communism and anything that might conceivably lead to it. Then the evils of the present system can even be admitted, and also accepted, as the alternative is so much worse. It is for such reasons that, though the twentieth century has been fundamentally one of popular progress, the atmosphere of social thought in the intellectual circles of the 'free' world has tended more and more to obscurantism and reaction.

ILLUSION, HYPOCRISY, BRUTALITY, DESPAIR

It has veered between the extremes of *illusion* and *hypocrisy* on one side, and *brutality* and *despair* on the other. Even the limited degree of consistency achieved by the free enterprise liberalism of the nineteenth century could no longer be sustained as the twentieth century advanced. Underlying this failure lay a basic contradiction. Either the indefinite continuation of the capitalist system had to be justified on rational and ethical grounds, according to the liberal tradition in which it had been formed, or it had to be maintained by force using mythical or religious sanctions, which effectively denied the very values which had made it work. Whenever conditions became really hard it was difficult to justify capitalism in the face of the common experience of its failure, and the knowledge of the existence of an alternative to it in socialism. At this point it was necessary to resort to force and fraud, but doing so under-mined the whole moral base for capitalism. The use of undisguised brutal and cynical force was the Fascist solution. Its ultimate failure showed that plausible deception offered much better protection to capitalism, especially as it could be operated through the whole new machine of a popular Press, cinema, broadcasting, and television in-cluding the pervasive effects of advertising, all firmly in the grip of big business or its representatives.

THE REIGN OF ILLUSION

The mechanism of democracy has under capitalism served well enough to maintain the rule of an almost hereditary minority, either in the gentlemanly English way, or in the more frankly corrupt and crooked ways of American or French politics. 'Democracy', as Bagehot defined it in his book on *The English Constitution*, 'is the way to give the people the greatest illusion of power while allowing them the smallest amount in reality.'[7.6] In the United States the coincidence of high wages and the carefully preserved political backwardness of the workers, together with a corrupt and progressive-proof representational system, enabled an illusory version of the liberal, free-enterprise philosophy to flourish for many years. Even there it has had to be reinforced by an anti-communist and loyalty campaign that aims at branding any opposition to the indefinite rule of capitalism as treason.

THE WELFARE STATE

In Britain, however, the working class, the first to be created in the modern world, were more politically experienced, and already by the thirties had lost most of the advantages that had accrued from having belonged to the workshop of the world. They could no longer be held by the illusion of liberalism, but their aspirations were met for a time by the ideal of evolutionary socialism or the Welfare State, which though more subtle, was to prove, in the era of the Cold War, as much of an illusion. The mass of the people could continue to believe that they were on the way to a new world of peace and social justice, while their betters in positions of authority and influence had arranged that things should go on as they were, and could explain the lack of progress by stressing the need for consolidation and caution so as not to compromise what had already been won. The ideology of reformist social democracy, particularly in its British 'Labour' form of a blend of Fabianism with right-wing trade unionism (pp. 1110 f.), did not in any case demand any basic criticism of the capitalist order, and emphatically repudiated the idea of any drastic change which might be called revolutionary. It was more comfortable to believe, or at least to pretend to believe, that the Welfare State had already arrived, or, if not quite, that it would come in due time given patience or restraint on the part of the 'under-privileged'.

The idea of the Welfare State in which the basic necessities of life are secured to all, the health of the people is protected, and there are full opportunities for education, has been the aspiration of generations of working men and women. It was largely to achieve it that they voted

339. The depression in the 1930s brought wide-scale unemployment. The Welsh, as well as those from other parts of the depressed areas, marched on London. Here a marcher is being given tea upon arrival in Cardiff in October 1936 prior to the march (compare with illustration 324).

the Labour Party into office in 1945. It was an objective worth fighting for, as is well shown by the persistence with which it has been attacked by reactionaries. The reality, has, however, proved very different from the ideal. As far as Britain is concerned, the benevolent State watching over and supporting all its sons and daughters from the cradle to the grave is a myth. The total of social services, provided by government funds, amount to 10 per cent of the National Income, about £25 per head per year, which is only three-quarters of its armament expenditure. For the benefits they get the workers themselves pay out in contributions another £10 a year and for the rest the money has been largely provided by them in indirect taxation. The quality of the services, the over-crowded schools, the non-existent health centres, the niggardly treat-ment of the pensioners, are what might be expected from an embarrassed charity organization rather than from a sovereign people dividing its wealth equitably among its citizens.

On the productive side the socialism of the Welfare State has been diluted beyond recognition. The industries that have been nationalized

340. The unemployed tried to scrape a living as best they could. Here they are entertaining a theatre queue in the West End of London, 1932.

are those that had already ceased to pay. They are run in such a way as to provide raw materials and services at the cheapest rate as an effective subsidy for the profitable parts of industry, which remain firmly in private hands.[7.17; 7.81; 7.100] The management of nationalized industry, despite all the past discussion about workers' control, is still firmly in the hands of the old directing class. It is not surprising that the enthusiasm that greeted the Welfare State had in six years evaporated sufficiently to allow the return of a Conservative Government. That Government maintained itself and even got re-elected, not so much as a maintainer of the Welfare State but of *full employment*. The illusion of full employment is a more subtle one. There is no question that it has existed for the best part of the nineteen years since the war. It is supposed to be the fruit of the application of Keynesian economics (p. 1134) to remove the recurrence of slumps (the credit for this achievement is disputed between the Labour and Conservative parties), but it may well have been due to a combination of external economic circumstances which in the present period of 'redundancy' seems to be showing signs of coming to an end. With it may go another illusion, that of capitalist stability. In any case the advantages of full employment were not all they seemed. According to Keynes' formula they were so well compensated by rises in price that a large proportion of industrial workers could only make ends meet by systematic overtime, throwing away all the advantages of the struggle for shorter hours.

The illusion is not, however, in the possibility of a Welfare State or full employment, but in the belief that they can be won in the face of capitalist opposition without a far more energetic fight than has yet been put up for it by the Labour Party. For in spite of the shortcomings of what has so far been achieved, the ideal of the Welfare State and the sense of power that the people get from it represent ultimately an irresistible political force. Once the people become enlightened through their own struggles and see what needs to be done they will be able to turn the illusion into reality. Economic democracy requires that the elected government controls all basic industries and services and that the privileged position of the present moneyed classes is definitely ended. Only then will it be possible for the whole of the material and human resources of the country to be used in the interests of the people.

HYPOCRISY

It is difficult to draw the line between honestly held illusions, which served to reconcile people to capitalism by making them feel that it

341. Overcrowding of schools in Britain after the Second World War brought strain and discontent among teachers in a country with 'full employment'. Photographed in Manchester by Bert Hardy.

could be improved or had already been improved out of all recognition, as might be inferred from the official publications of the Labour Party,[7.88-90] and downright hypocrisy designed to cover up its evils under high-sounding phrases. One of the most effective ways of distracting attention from the hardships and anxieties of the present is to show them as lesser evils as compared with an alternative régime that would make everything much worse. For this anti-communism has served for many years, but never more effectively than in the early years of the Cold War.

Between 1917 and 1947 a complete change had come in the stock-in-trade of phrases used in every organ of publicity to attack communism or justify capitalism. At first it was in the name of defence of property, order, and tradition that right-thinking men were urged to defend themselves against red revolution, anarchy, and violence; by now it is in defence of liberty and free institutions that the same papers demand the breaking-up of totalitarianism, the police State, and red imperialism. As a tribute of vice to virtue, this hypocritical switch-over indicates the enormous growth of popular strength and the abandonment of the old reactionary positions. People who never opposed and often helped the Nazis have made a most clever use of the contempt and hate in which they came to be held. Their sins have been transferred in totality on to the government and people that were their only resolute and effective opponents.

Even to the extent that the criticisms of Soviet rule were justified in the later Stalin period (p. 1172), they were never, on the part of those most vociferous in pressing them, the real reasons for their hostility, as is shown by their keeping it up after most of the evils they complained about have been removed. This hypocrisy reached its height in the Cold War, and especially in the citing of China by the United Nations as an aggressor in Korea.

These are mere examples, however, of the underlying basic hypocrisy of our time, in which the wealthy and privileged, without surrendering a jot of their advantages, pose, and largely pose successfully, as the true friends and supporters of the aspirations of the very people on whose exploitation they have grown fat. It is not surprising that in political affairs words like democracy, security, defence, and peace have all but lost their meaning. It was in the First Great War itself that this practice of deception first reached massive proportions. Gentlemen who were making good profits and who held the people in contempt publicized it as a war to end war and to make the world safe for democracy. As these fine phrases were soon found to be empty, a general cynicism took the place of wartime idealism and opened the way to Fascism.

THE REIGN OF BRUTALITY:
THE NAZIS AND AFTER

After the First World War, it gradually became evident that there was no prospect of the attainment of prosperity for all under capitalism, but only enough for a privileged few. This realization came first in the less successful or defeated capitalist countries, then in the others, and, with the onset of the great slump, even in the United States. As this was recognized, more and more people, especially middle-class or would-be middle-class, began to listen to the violent, ignorant, and unbalanced men who had been claiming that one race or culture had the right to dominate the world by force. We have been led to associate this ideology, with its open contempt for reason, morality, justice, or democracy, first with Italian Fascism, then with German Nazism, the political products of the first and second world slumps. It is in fact much more pervasive as a symptom of the consciousness of the decay of capitalism, and shows itself in many ways, of which anti-Semitism and the colour bar are only the most obvious.[7.16]

Brutality and exploitation had always gone with capitalism, but they used to be concealed behind a façade of liberal democracy. Their open emergence was a sign that the capitalists felt that liberalism and democracy were luxuries that it was no longer possible to afford. The real backers of Fascism and Nazism, the big bankers and industrialists of Italy and Germany, were no different from their colleagues in America, Britain, and France, and co-operated closely with them up to and even after 1939, when they still hoped to use the Nazi forces against the Soviet Union.[7.180] They are now boasting that throughout the War they never abandoned this hope, but had to keep it in cold storage for the day when they could again unite with the same Germans in defence of Western Civilization.

Events showed, however, that brutality in its crudest Nazi form did not, in fact, provide the solution. Instead it aroused such world-wide popular opposition that when the Nazi State was smashed in the Second World War, after immense sacrifices, by the power of the Soviet Union, there was a danger that its fall might endanger the capitalist system itself, at least in Europe. In Asia the defeat of Japanese imperialism actually had that effect in almost every country it had occupied.

THE DOCTRINE OF PEACE THROUGH STRENGTH

Nevertheless, the conditions that led to Nazi brutality still exist, and in a different form it is still vigorously in action. Under a covering of hypocrisy, and with the help of an immense propaganda machine, the old idea of salvation through force has been revived. The real objective remains unchanged. It is still that of restoring the world control of big business through the use of an overwhelming military force, equipped by the immense technical capacity of the United States. It is for this end that the latest scientific techniques – bacterial warfare, atom and hydrogen bombs – are being frantically developed. The idea is the same: domination through destruction, whether it is done in the name of a Master Race or Western Civilization, 'Democracy', and a 'Free World'. It has been further shown by the cruel and futile wars in Algeria and Kenya and by the struggles still going on in Angola and Rhodesia, the last bastions of the old colonial world.

The policy of brutality relies for the most part on simple appeals to aggressiveness, self-interest, or fear. It has little new to offer to social thinking. Indeed it is essentially an anti-intellectual, know-nothing movement, rapidly becoming subliterate, drawing its inspiration from comics, the cinema and commercial television, with their glorification of brutality, sadism, and scientific slaughter. For justification it relies on traditions and on a sentimental and obscurantist religion. Here the more reactionary elements of the Catholic Church have raised the doctrine of anti-Communism to the status of a new crusade, in which anything is permissible to destroy the influence of the Evil One.

On the scientific side the stock ideas of Fascism – race and anti-Semitism – were found to survive the defeat of Hitler. Though discredited in their crude form, they are still practised in the southern states of the USA and in the Union of South Africa, and they have taken new life in the doctrines of neo-Malthusianism, already discussed (p. 973). This idea of an overcrowded planet is taken to justify the doctrine that the backward peoples, who by their very existence threaten the more fortunate heirs of Western Civilization, must be checked, allowed to die of disease, or at least prevented from breeding.[7.172a]

THE PHILOSOPHY OF DESPAIR

There were many, particularly among the intellectuals in capitalist countries, who were too clear-headed to accept the illusions of liberalism and social democracy and were too humane to fall in with the doctrines of brutality and destruction. Under the impact of the tragic events of the century some of these, a small minority in some countries, an important

342. Personal involvement was part of the doctrine of existentialism. Its chief protagonist Jean-Paul Sartre wrote a script for the film *Witches of Salem* in which this sense of involvement was stressed.

one in others, broke altogether with the system, in which they saw no hope, and joined in the struggle against it. For those who had neither the desire nor the courage to do this all that was left was to retire from the world of action and to adopt attitudes of cynical or melancholy despair, as had their predecessors at the time of the fall of the Roman Empire. This tendency to escape became especially marked in and after the Second World War. Its simplest form was that of absorption in scientific or artistic work, with a more and more deliberate ignoring of political events and avoidance of political action. In the more extreme cases it has led to the abandonment of the whole intellectual advance of the Renaissance, with its emphasis on unlimited horizons for humanity, and a return to medieval or oriental mysticism (p. 301), sometimes to incongruous mixtures of both. In some of its aspects *Existentialism*, with its emphasis on guilt and anxiety, also expresses the more intellectual side of the philosophy of despair. However, it contained, especially in the person of its chief protagonist Sartre, another element drawn from the resistance movement, the *engagement* or personal involvement that can give it once again a positive content.

WESTERN CHRISTIAN CIVILIZATION

Of all the reactionary conceptions that have flourished since the Second World War the idea of Western Christian Civilization is the most ingenious, plausible, and dangerous. Like the Holy Alliance of 1815, of which it is the modern counterpart, but unlike the crude anti-Comintern Axis, which had exactly the same ends, it combines an appeal to respectability, religion, and tradition. It has also, with the Western label, an implication of modernity and superiority over Asiatics and other lesser breeds.

As an historical or social concept it does not bear examination. Despite the efforts of world historians, like Toynbee[7.164] and Spengler (1880-1936),[7.153] with their mythical cycles, the real equation of Western Christian Civilization with monopoly capitalism, and particularly with Wall Street, cannot be seriously disputed. Certainly those intellectuals who take part in this great enterprise know well enough who is paying the piper. Yet as a rallying point for a crusade against communism the flag of Western Christian Civilization still has its advantages. It can be made to cover the most unsavoury allies, from Franco to the late Ngo Din Diem. For such a noble cause, who would worry about the use of such necessary and just means as the hydrogen bomb, that great deterrent, even if it means killing half the population of the world, including all those of an expendable Europe?

From this one overwhelming example we can see that the tendencies of illusion, hypocrisy, brutality, and despair are not distinct. One merges imperceptibly into another, and all are symptoms, more or less consciously recognized, of the failure to react positively and constructively to the social transformation of our time. They all stem from an instinctive, but by no means disinterested, attachment to the advantages that class society can give to those willing and able to profit by it. They all imply a basic contempt for the common people, or at best a lack of faith in their capacity to build a new life by their own efforts.

The sketch given in the preceding paragraphs is aimed at providing the relevant background against which the social sciences in the capitalist world have developed in the last forty years. The stress of general events and the strains they have exerted on the individuals and institutions go a long way to explain why the social sciences, far from moving towards some new synthesis, present many confused, incoherent, and often retrogressive tendencies.

13.2 The Social Sciences in the Capitalist World

Any critical study of the present state of the social sciences would show very clearly how progressively disintegrating have been the effects of the intellectual and social atmosphere of capitalism. There has been, it is true, a great increase of interest in the social sciences, and money, little enough compared with that spent on the natural sciences but on a scale hitherto unheard of, has been spent on furthering them. Hundreds of serious and, for the most part, sincere and well meaning workers have been engaged in social research. Nevertheless, in spite of a vast volume of detailed studies and surveys, no radically new theory has emerged. What has occurred in the last fifty or more years, especially in the last twenty, is that a new twist has been given to old theories to help them to explain or rather to explain away the failure of capitalism to live up to its old promises.

Maynard Keynes (1883–1946) sought to prove that economic crises were due purely to financial maladjustments, technical hitches in the beautifully balanced machine of capitalism. He explained that all that was necessary to cure crises was government control of investment policy and the freezing or reduction of real wages, which would secure full employment without any change in a profit-making economy. Freud and the psycho-analysts explained that the drift towards war was due to inherent unresolved aggressive instincts. Political action was useless or harmful until these had been made harmless by analysis, and as this patently could never be done we had to put up with things as well as we could. Positivists demonstrated on the other hand that all our social troubles were due to the misuse of language. If all politicians could only be convinced that what they were saying was intrinsic nonsense they would hand over to semantically trained social scientists.

These views, which are by no means exaggerated from the originals, show to what a pitch of intellectual bankruptcy the social scientists had been reduced even before the Second World War. It is not surprising that their theories were impotent as bases for effective action in the affairs of the world. They had, however, other uses: those of providing an intellectual camouflage for capitalism and the means for making it run more smoothly. Though the last forty years' work has produced no great theoretical advance it has witnessed an enormous growth in the techniques of the social sciences, especially in the greater use of statistics in the methods of questionnaire and survey and in the analytic methods of operational research.

In the ensuing pages will be found a brief account of the development of different branches of social science, notably in economics, political science, sociology, education, and lastly in the most ideological fields of all, in psychology, philosophy, and theology. Nothing further is said here on the historical sciences, which have been treated in the previous chapter (pp. 1079 ff.).

THE KEYNESIAN REVOLUTION IN ECONOMICS

The most characteristic of recent modifications in the social sciences has been that effected by Maynard Keynes in economics.[7.43] Its importance has been exaggerated both by the supporters of the older school of economics, who consider that he has thrown away the whole case of unfettered private enterprise, and by Labour leaders, who hail him as justifying Welfare State socialism. Actually he made the minimum alteration to the marginal economic theory (pp. 1087 ff.) which would give it any chance of corresponding to reality in the world of monopoly capitalism. What he sought to prove was that the capitalist system does not necessarily, by itself, lead to a condition of full employment. This idea, however horrifying to a classical economist, must have seemed somewhat of an understatement to the millions of unemployed at the time it was written.

The solution he proposed in *The Theory of Employment, Interest, and Money* was far from that of abolishing the capitalist system, to which he was deeply attached. It was rather a question of mitigating its effects by State action on investment, together with a manipulation of prices aimed at keeping down real wages. In this he recognized, somewhat belatedly, that since the First World War, as Lenin pointed out, 'monopoly capitalism had passed into State monopoly capitalism'. The line between State and monopoly interest had become very difficult to define. In time of war the connexion was openly admitted; government controls operated through the agents of the monopolies. When, in response to working-class agitations, certain industries and services had to be nationalized, they were restricted to practically bankrupt ones, and their management was left substantially in the same hands as before.[7.81]

Of Keynes' remedies for slumps, that of useful investment in long-term capital projects was never in fact employed. Nor is it likely to be, for it demands greater government expenditure just at the time when every moneyed interest is howling for economy. What halted the first depression was the movement to rearmament, which Hitler began, and since the War full employment has been maintained by the same means. This is the only way by which capitalism can restrain, for a time and at terrible human cost, its tendency to produce more than can be consumed by a population deficient in purchasing power.

The other Keynesian remedy, the policy of wage freezing, was anything but academic. As operated in the War it consisted of holding wages down while adjusting the prices of basic foods – the iron ration – by subsidies which cost far less than would have wage increases proportional to the rise in the real cost of living. Later it was operated loyally by a Labour Government in Britain – to its own destruction – fully supported by some trade union chiefs, and was only abandoned under strong popular pressure. It was taken up again by the Conservative Government in 1960 and had only the result of slowing down British production below the level of that of other capitalist countries of Europe.

The value of Keynes' work to the rulers of Britain and America lies in the fact that it seems to supply a scientific and impartial economic analysis, neatly reconciling the interests of employers and labour in a period of capitalist decay, when the exercise of unfettered competition could only provoke revolution. The success of the theory with the bi-partisan leaders of Western Democracy is an excellent example of the use of a social science as a cloak for class domination. As Lenin said:

> To expect science to be impartial in a wage-slave society is as silly and naïve as to expect impartiality from manufacturers on the question whether workers' wages should be increased by decreasing the profits of capital.[7.93]

The new economics of harmonization and restraint acts in a very one-sided way, fettering trade union activity while giving employers full liberty. The form that it has taken in recent years is the productivity drive, which has attractive plausibility.[7.79] It implies that the amount of production is determined solely by the workers' willingness to accept more strenuous working methods or to work longer hours, whereas it is much more dependent – given the fact that the employer controls the conditions – on the capital available, and ultimately on financial considerations. The greater productivity of American workers is admittedly due almost entirely to the better equipment they use.[7.1] Yet an English manufacturer using equipment which is obsolete, but which has paid for itself over and over again, still contrives to make a good profit.

Such questions were hardly treated by Keynes himself, who worked on the assumption of a technically static economy. His most distinguished pupil, Joan Robinson, has now extended his doctrines to cover the conditions of monopoly capitalism in a world of rapid technical change. In *The Accumulation of Capital* [7.138] she treats theoretically of the conditions under which it is strictly profitable to improve technical methods, taking into account costs, obsolescence, and prevailing wages. It seems evident from her analysis that the monopolist capitalist system becomes progressively more incompetent to provide optimal increase in welfare as technology advances, for this introduces an additional element of uncertainty and has a discouraging effect on investment. Here again the economist demonstrates, somewhat after it has become apparent, the bankruptcy of capitalism and the necessity for a socialist alternative.

POLITICAL SCIENCE

Before 1914 political science travelled more or less comfortably along the lines prescribed by liberal progress. Outside the Marxist camp it was assumed that the ideal, which would sooner or later be reached all over the world, was one of parliamentary democracy, guaranteeing property and leaving the field free for private enterprise. The events of the succeeding years showed this to be an empty dream; parliamentary democracy did not spread, indeed it lost ground everywhere. Even when the interventionary wars and the *cordon sanitaire* checked the spread of Bolshevism the result was not more democracy, but the establishment of near-Fascist or Fascist dictatorships.

Even in the old centres of democracy the political actions of parliaments began to be dominated by economic interests. The State could no longer keep the ring of the economic prize-fight; it had to intervene actively to direct or control the internal and external economy of the country. In the arch-individualist United States itself the impact of the slump forced the acceptance of the New Deal. This was essentially an admission of government responsibility for the welfare of the citizens, who could no longer be expected to cope for themselves in the face of economic changes of world-wide range.

It might have been expected that these great changes, with the explosive influence of science on technology, the increasing integration of State and private concerns, the merging of political and economic man, would have led to a rapid growth of political science and the formulation of wholly new principles. This, however, did not occur. Such changes of views as took place were reactionary, echoing a religious or Fascist authoritarianism.

A notable example of this kind of political thought has been the thesis of *The Managerial Revolution*, put forward by James Burnham in 1941.[7.19] His subsequent work included a gleeful anticipation of the crushing of the Soviet Union by Hitler, and as that failed, a desperate advocacy of American supremacy by military force and instigated rebellions.[7.18] His thesis is that there is no question of a transition from capitalism to socialism, but to a *managerial society*, where effective control has been peacefully and imperceptibly usurped by a new managerial class composed of foremen, managers, and technicians, who, thanks to the growth of modern scientific industry, have become indispensable to the modern State. World history does not, according to Burnham, lead to revolution, but to wars for supremacy between giant managerial States. These, after the defeat of his favourite Germany, have been reduced to the USA and the USSR, of which the victory of the first is assured by its economic and technical superiority. Here a pseudo-scientific veneer has been thrown over the naked *geopolitik* of the German militarists. The managerial idea has been most rapturously received in the Press of the great monopolists whose supersession it announces. They are well content to let people believe that their rulers are the salaried staff who work obediently for the monopolists, and are occasionally elevated by blocks of shares or seats on directors' boards to a minor position in the hierarchy of wealth.[7.115] It has also obtained considerable support among the Labour and social democratic intelligentsia, who are flattered to find themselves, without any effort on their part, as a new ruling class. In fact, nothing has changed in the domination of capitalism except a tighter control by the financial magnates of the big corporations, as shown in the American witch-hunts, where any criticism of the rule of the wealthy is stigmatized as communism.

Apart from such extravagances as Burnham, and the depressing fatalism of Professor Toynbee's mortal civilizations (p. 1241), the official exponents of political science continued, for the most part, to reiterate in more and more pained and strident tones the commonplaces of the nineteenth century, disregarding the fact that they had little bearing on the contemporary world.

Even when, as in Britain, popular feeling brought the Labour Party into power, no socialist theory guided its actions. Indeed its leaders prided themselves on their adherence to scientific, that is capitalist, political theory, and abided by rules which had been made by their enemies to prevent any significant change in the pattern of distribution of wealth and power. In fact, the main use they made of political theory was to prove that with the reforms they had instituted in their first terms of power everything not only possible but also desirable had been done. The 'silent revolution' had, they contended, in fact succeeded without, however, attracting much attention. The ideas of socialism were lifted on to a higher plane.

This is the theme of much recent Labour Party publication, notably *20th Century Socialism*,[7.150] which, as a way of celebrating Keir Hardie's centenary, marks an absolute rejection of his definition of socialism as 'collective ownership of all the means of production, distribution and exchange'. However, there are signs that a new wind is blowing. Political thought is bound to be affected by the increasing economic strains of capitalism and by the movement towards democracy in the Soviet Union. Political science can find again its true purpose as an agent of social understanding and change. The achievements of the Soviet Union, as witnessed by the sputniks, have brought political thinkers in capitalist countries to realize that they are in a new world of science and technology. This can be seen by the concerted drive to introduce new policies in education and research, although it has not yet led to any large appreciation of the social factors of the situation (p. 1153).

13.3 The Application of Social Science

Though the more academic aspects of social science, such as the theories of economics and politics, have found their main use in the ideological defence of capitalism, the great growth of the social sciences, especially since the Second World War, has been due to the realization of the cash value that their application can bring in actual working of the capitalist system. Social scientists had already before the War found employment in economic surveys, in market research and advertising, in the promotion of harmonious industrial relations, in urban and regional planning, and in education. In order to be able to do anything useful, or even to appear to do so, in all these fields something more than the vague generalities of academic social science was needed. It is indeed striking how little the theory of sociology has advanced since the period of the First World War. The early brilliance of Hobhouse, Wallas, or Tawney in Britain, or of Veblen, Weber, and Dewey in America, has not been followed up. Instead a profuse and windy dullness has settled on the subject, as anyone can verify for himself by trying to read works published in the last twenty years. Official sociological theories are so highly formal and emptied of any trace of historical development that it is difficult to see how their authors manage to extract from them a justification of all the commonplaces of reaction.

QUANTITATIVE AND STATISTICAL METHODS IN THE SOCIAL SCIENCES

There has indeed in recent years appeared and grown amazingly a very different kind of sociology, in which measurement and calculation have taken the place of verbal definition and generalization. This development gained an enormous impetus from the events of the Second World War, though the effects were not to be as lasting as was originally thought. The War made it possible, for the first time, in such countries as Britain and to a lesser extent in the United States, to see something of what applied social science would be like. The sociologists, who had hitherto occupied themselves with academic studies, found themselves in a situation where they were called on to do something as the result of their investigations and, what is more, were given the means both to make the investigations and to have the conclusions put into practice. They also found themselves working side by side with physicists and biologists in the field of military operations and industrial production. The effect of working together was very beneficial to both social and physical scientists. The physical scientists learnt the advantages of the techniques of survey and statistical analysis that had been developed by the social scientists. The latter in turn learnt the value of planned experiment and analysis of variation that were the stock-in-trade of biological and physical scientists (pp. 833 f.).[7.11.285]

The most powerful and general of the new methods was the statistical. Whatever names they go under – social survey, opinion polls, social and industrial psychology, market and operational research – they all consist essentially of a more or less statistical analysis of data about thinking, working, or living situations extracted by systematic inquiry.

Statistics had long been used in economics and government affairs, but essentially as a means of recording and summarizing facts, as in a budget or a population census. Their new application as a research tool for asking and answering general questions came in the first place from the biological field, largely through Karl Pearson (1857–1936), himself influenced by the eugenics of Galton and the positivism of Mach. His *biometrics*, with its practical application to agricultural crop yields, gave rise to sociometrics, which attempted to weigh attitudes and opinions, even intelligence itself. Mathematical methods of extreme elegance and precision were devised which showed how much could or could not be learned from a limited series of highly irregular and unreliable facts.[7.58]

SOCIAL SURVEY

As long as these data were material quantities, and were honestly gathered and analysed, the results could be of real social value. The tradition of Booth and the Fabians of 'measurement and publicity' (p. 1097) has been carried on with an increased range and penetration into the twentieth century. One effective form it has taken has been that of social survey, where, by using samples, a quantitative picture can be gained of the over-all state of a particular area or group. In Britain[7.120] the nutrition surveys of the thirties, already referred to (pp. 878, 898), which linked food consumption with income and health, revealed that in the most prosperous country of Europe 50 per cent of the adults and 25 per cent of the children were undernourished. Their publication paved the way to a popular agitation that forced at least some small concessions. With similar studies in housing and education the result has been to arm those demanding improvement with factual data on which to base their claims, and to provide a quantitative yardstick to measure any degree of improvement secured. I have myself some experience in factual social surveys.

I helped to plan and direct a large-scale survey in two British cities during the War, in which the object was to relate intensity of air attack with production. We measured everything from the consumption of beer and aspirins to the production of machine-gun bullets. It was the first and, I believe, the only time that the social and productive data on a modern community have been related to each other. We did not find the conclusions about the effects of raids that our taskmasters desired, so they were obliged to carry on in defiance of the findings; but we did find some most interesting relationships between work and wages of which no use was made. We found, for instance, that the rate of wages paid had much more influence on drawing workers into the town than the bombs dropped had on driving them away. It was also found, and this was not so acceptable to the authorities, that, if the workers in Birmingham had been concentrated in the factories with the most recent machinery, it would be possible by working several shifts to produce twice the amount of war material from Birmingham without adding a single man or machine. The existence of several hundred small and ill-equipped metal firms in Birmingham fulfilled the economic function of establishing a level of cost which enabled the big firms to make very large profits during the war.

Especially during the thirties the social survey became, not, as the Fabians had intended and hoped, a means of persuading the ruling classes to extend assistance, but a weapon used by the working class in agitating to demand and secure its rights. As time went on means were found to put a stop to this, not so much by suppressing surveys, but by subtly changing their character; the statistical method lived up to its reputation of 'lies, damn lies, and statistics'. National income figures and cost-of-living indices, painstakingly drawn up, showed results that people knew were wrong without being able to prove it. Abraham Lincoln's warning that 'you cannot fool all of the people all of the time' was borne out in the popular distrust of official statistics. Social and economic surveys themselves began to wear a new look. In the third Rowntree survey of poverty in York, the statisticians succeeded in virtually abolishing poverty on paper by ingenious devices which effectively pushed the standard far lower than it had been in previous surveys. This hand, however, has been overplayed. The plight of the aged pensioners and the low real wages of the majority of workers are beginning to provide even statistical expression of real and remediable poverty.[7.101]

OPINION POLLS

Manipulable results of surveys could be obtained even more easily where the basis of all the calculations no longer lay in the material world, but in an ideal one where everything was subject to opinion, and where unconscious bias could determine the answer despite all the refinements of calculation. This was particularly the case in opinion polls, whose use spread from America in the thirties. These suffer from the double disability that the type of answers depend on the slant given to the questions, and that they at best can only provide information about what people feel they ought to say, rather than what they really think. For the purposes for which they are used this is not necessarily a disadvantage. A poll can usually be relied on to provide the result the sponsor has ordered. If it palpably does not it can be played down or suppressed. Applied to politics, opinion polls are an insidious danger to democracy. An election is an action of the people: they express their will and it is acted on. A sample poll implies no popular power. It gives the same weight to the thought-out conclusions of a man of integrity and influence as to that of someone who merely does not want to be in the 'do not know' class. It is merely an indication to the manipulators of opinion that it may be necessary to change their tune. The people have ceased to be masters and have become a docile herd to be driven this way and that by the shouts and lures of publicity.

MARKET RESEARCH AND ADVERTISING

The greatest use of the method of opinion sampling is in so-called market research, where it has replaced, or at least supplemented, the opinions and hunches of the salesmen. It is part of the great apparatus of unloading unwanted goods on necessarily uninformed buyers, a process of increasing difficulty in the general crisis of capitalism. Here

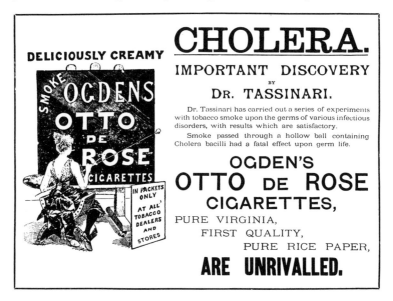

343. Advertising can provide a 'slant' that may not be wholly desirable or scientifically correct. This advertisement was published in 1892 in the *Illustrated London News*, the very year when a serious cholera epidemic swept through Hamburg and while there were fears of it reaching England.

the statistical method has the advantage in that it provides the result in numbers: a language which business men can understand. What the figures mean, or whether they really mean anything at all, is not the point. What matters is that they are of use in board-room arguments so that some people are willing to pay to get them, and consequently it pays to have people trained to produce them.

Whether expressed in figures or in the wordy generalizations of the social scientists, the purpose of the information given is the same, to assist in the making of profits by increasing sales. The twentieth century has witnessed the growth of a vast apparatus of advertisement and publicity which, by its very volume, blinds and deafens the majority of the people to any real appreciation of the world in which they live.[7·121] Advertisement has indeed become a vast parasitic industry, whose function is effectively to exact a very concealed purchase tax from the consumers for the benefit of a few newspaper and advertisement magnates. The tax is a substantial one, graded according to the worthlessness of the product and the poverty of the purchasers.

In a profession where the value of everything is what it can be sold for, the conclusions of social science have to be weighed against the rhetoric and display of the art of salesmanship and advertisement. In so far as science survives at all in this atmosphere, it has to bow to these requirements and be corrupted both in what it states and what it suppresses. Figures are as flexible as words, and more deceptive because they carry with them the air of neutral facts.

SOCIAL SCIENCE IN PRODUCTION

The use of social science in the processes of capitalist production itself is a more recent development. Except for the work of some self-styled scientific management experts, it dates from the last two decades. In its older form of time and motion study it was crudely designed to rationalize the motions of the workers in order to get more work out of them in the same time. It dehumanized men and women to the level of the machines they served. It was accordingly resented and resisted, with varying success, by the organized workers. The attempted introduction of such a system into a factory was often enough to provoke a strike. As one trade unionist of the old school remarked, 'Time and motion studies mean that time is lost and motion ceases.'

The later forms of application of social science to production are more subtle and ostensibly inoffensive. Two lines of approach have been followed: one beginning with the mechanical techniques themselves, and there concerning itself with the adaptation of the workers to the machines – here time and motion studies have been extended and glorified with the war prestige of *operational research*; the other approach starts with the human element, and strives by psychological methods to make him not merely a more efficient but a more willing tool – this is the province of *industrial psychology* and the *science of management*.

OPERATIONAL RESEARCH IN INDUSTRY

The genesis of operational research in relation to war needs has already been discussed (pp. 833 f.). Immediately following it there was a drive among both natural and social scientists to use the methods of operational research in the fields in which they were most needed, that is, in advancing agricultural and industrial production, and in social services like housing and health.[7.11.201] As has already been pointed out, the advances of physical and biological science now make it possible, in principle, to develop increasingly rational types of production in both spheres, and to achieve enormous economy of effort on the one hand and much greater satisfaction to the user on the other.

After ten years, however, the application of operational research has grown with disappointing slowness. In the original sense of its definition as 'the use of quantitative analysis of real situations as a guide to executive action', it has not fitted readily into the capitalist pattern. This is because industrial production has neither the unquestioning discipline of an army nor the sense of common purpose of a team or co-operative enterprise. Consequently, the method of operational research, with all its statistical refinements such as linear and concave programming, can be used only in a purely technical sphere or as a means of getting the most work out of employees, merging with the older and highly suspect scientific management. Nor even for these purposes has there been much enthusiasm, for in business circles it seems to infringe on the powers of managers. To make full use of this important method, which bridges social science and technology, would require really effective workers' control of industry and the elimination of private control and private profit.

This does not preclude a limited application of operational research to the strictly limited problems of getting the most work out of a factory or maximalizing profits. It has indeed been annexed to scientific management, and in the process has lost its independent normative status.

COMPUTERS AND SOCIOLOGICAL RESEARCH

It is in this field of operationaî research that the newly developed computers (pp. 783 f.) have found their greatest utility. Many of the judgements made in operational research on the basis of statistics can be made practically automatically by applying these statistics directly to the machines, setting them to compute all the different results that would follow from the different permutations of the policies adopted and choosing the most economical result. This is a method which is now only just starting, but it is bound to have great and even revolutionary effects on all human affairs. It has not found very much use yet in the capitalist world, owing to the fact that most business executives do not understand it and if they did they would object to it even more strongly because it trenches on their *raison d'être*, the so-called managerial function. It is for judgements of this order that the intuition of the captains of industry was deemed to be indispensable. In the Soviet Union, however, there is no such inhibition; as computational methods improve they will become universally adopted. It is not only in imitating human judgements that the machines are going to be valuable; they will gradually and then more and more rapidly take the place of the economic arrangements themselves – of banking, credit, ultimate

344. Analysis of the results of an opinion poll.

money payments in wages and purchases. In fact, they offer the possibility of mechanizing the whole of the economic system.

INDUSTRIAL PSYCHOLOGY

There remained, however, considerable scope for another kind of social science, one which could be relied on not to interfere but to smooth the way of capitalism. This was provided by a development of the methods of industrial psychology and scientific management. Here the prime object is to control the mind and will of the worker by the use of every device developed by social or psychological science.

In the United States this development dates from the days of the slump; in Britain it is more recent, but in both cases it received an enormous impetus from the Second World War. The new prestige of psychology arose largely out of the claims for its successful use in the selection and training of the armed forces. It is hoped that similar methods may help to provide even better-disciplined industrial armies. The underlying aim of industrial psychology is to secure the maximum will to work and to create the impression of harmony of interest between the workers and the management.

345. The problems of industrial psychology are increasing with the use of automation, as here where machine tools are electronically programmed and obviate the need for human operatives at each machine.

The implication of the term psychology, with its psycho-analytic and psychiatric overtones, is that most workers – slackers, absentees, and, worst of all, agitators and strikers, any, indeed, who rebel against exploitation; in fact all but the model – are sick people requiring diagnosis and treatment to adjust them to their conditions of work. Their home lives need to be looked into, their wrong ideas about the owners corrected, their uncooperativeness modified by group exercises. Everything indeed needs to be done for them except the one thing from which they would really benefit – the end of exploitation. As long as exploitation of workers for profit remains, the only result of successful industrial psychology can be to hinder the worker from pressing his demands for better wages and conditions, and thus to diminish his share in the product of his labour. To use science for this purpose is to make it accessory to deception, even when the scientist himself is not aware that he is doing so.

THE CORRUPTION OF SOCIAL SCIENCE

The social scientists working in these fields may genuinely believe they are working for the good of the community in promoting social harmony, or they may hope that in the course of their work they may add to human knowledge, or, more cynically, they may find it a good paying job. The enterprise is liable in all cases to be scientifically sterile because it is limited in advance by the conditions of its use. The double need to produce practical results satisfying to the employers and to give the appearance of scientific impartiality, so as not to upset the workers, prevents any serious analysis. What appears instead is only too apt to be a mixture of impressive statistics and pronouncements of the most obvious platitudes expressed in pompous scientific language based on questionnaires full of questions of no importance, casually or inaccurately answered. An extreme but true example is that of an American industrial psychologist who discovered that a miner identified the mine with his mother and the boss with his father, making him unduly gentle in his work and aggressive in demanding his rights. He hoped that by changing the roles the miner could attack the coal face with vigour and turn into a model employee.

The full value of a worker as a human being as well as a producing unit can be realized only in an economic setting in which he is master of his work. This, limited as it was by poverty and ignorance, had been the condition of the medieval craftsman. It can be recovered in a world of scientific industry only by a type of organization in which the individual worker, in conjunction with his workmates and with technical advice, really determines how the work should be done, controls and directs the machinery, and reaps the full benefit. Anything less is deception and will sooner or later be detected.

Trade unions, until very recent times, have always fought against all forms of 'scientific management'. But this attitude has for some time been considered old-fashioned by the leaders of American trade unions, who even hire industrial psychologists to assist firms to become more efficient. For a while in Britain, at the height of the Cold War boom, similar moves to improve productivity by psychological methods gained support from a few powerful trades union leaders who apparently considered that the defence of democracy is best secured by preserving and strengthening capitalist monopolies. Mutual understanding between men and management in the interests of production still enjoys the support of Moral Rearmament and even some discreet royal patronage. It has, however, fallen into some disfavour among trade unionists, since it is becoming apparent that with higher productivity they are simply pushing themselves out of jobs. Nevertheless, on account of rearmament, capital investment and industrial research are curtailed (pp. 843 f., 1257 f.), and the boosting of industrial psychology, which is comparatively cheap, is the only resource left.

These considerations go some way to account for the markedly more favourable attitude in recent years of authority on both sides of the Atlantic towards the social sciences. They had shown value and they knew their place. Their role was to be that of a servant, not of a counsellor: 'on tap but not on top', as an Air Vice-Marshal elegantly put it.

Social science teaching in the universities still remains devoted to its old pre-war tasks of justifying the existing capitalist, or slightly modified capitalist, form of society. Nevertheless the idea that social science should be taught in the schools was quite rightly rejected by the Norwood Committee in 1941.[7.34] However well it might be presented, it was felt that it might lead to the spread of dangerous thoughts.

13.4 The Science of Education

Standing somewhat apart from the other social sciences, and even less sure of its scientific status, is the science of education. Ideally this should cover the whole processes of conditioning, ranging from birth to death, by which human beings are adapted to the society in which they live and the means by which they learn to make the best of it or to change it. Actually, as a study, it has arisen very tardily in our own time out of the practical difficulties of a school system attempting with grossly inadequate means to provide for rapidly growing educational needs. The demand for popular education, as apart from specialized training for church, law, or medicine, first arose with the insurgent bourgeoisie of the sixteenth century, and was closely connected with the allied movements of religious reform and political liberty. Popular education had to fight for recognition in the seventeenth and eighteenth centuries; that is why the early educators, like Vives (1492–1540), Comenius, Rousseau, and Pestalozzi (1746–1827), were also notable philosophers and reformers and played a large part in overthrowing the ideology of the feudal order (pp. 451, 1055).

CLASS DISCRIMINATION

However, once industrial capitalists were in power their enthusiasm for the extension of education soon evaporated. True, the new working class needed enough acquaintance with the three R's to do their jobs properly, and provision for teaching them was reluctantly provided on the cheapest possible basis. But there was all the more reason for seeing that the education of the masses did not go too far, and that it introduced no unsettling ideas.

346. Students at Ruskin College, 1906.

Conservative opposition to education, and particularly to higher education of the workers, was firmly maintained. When, in a mood of liberal emancipation, the London Mechanics Institution (later Birkbeck College) was set up in 1823 under the inspiration of Thomas Hodgskin (p. 1066), with the stated object of providing information to working men on 'the facts of chemistry and mechanical philosophy and of the science of the creation and distribution of wealth', the *St James Chronicle* wrote:

> A scheme more completely adapted to the destruction of this Empire could not have been invented by the author of evil himself . . . every step which they take in setting up the labourers as a separate and independent class is a step taken, and a long one too, to that fatal result.

Over most of the continent of Europe this danger was avoided by virtually closing access to higher education to all but a practically hereditary educated class, the so-called *intelligentsia*, who, with a few brilliant exceptions, furnished all the clerics, lawyers, doctors, scientists, engineers, and administrators that the countries needed. This rigid system of class-limited education was first broken down in Russia after the Revolution, amid bitter protests from the more reactionary of the old intelligentsia. It has now given way not only there but all over eastern Europe and in China to a fully popular educational system of a radically new kind (pp. 1187 ff.).

In England – not in Britain, for Scotland has an old-established tradition of popular education – where industry was first to develop on a large scale, the arrangements were somewhat more flexible. Just sufficient entry was permitted into higher education to meet the needs of a rapidly expanding production and administration. But the entry was so regulated that the new recruits could be assimilated into the ranks of the governing classes and often became their staunchest supporters. The advance of popular forces has widened the range of access, but the fundamental principle of upper-class assimilation remains.

In the new countries, particularly in the United States, education was from the beginning relatively cheap and easily accessible. At the same time its quality was low, and it counted for little compared with practical ability and business sense. The early achievements of American industry were due to practically illiterate inventors, which only goes to show how much more important, at that primitive technical stage, were favourable economic conditions and good craftsmanship than any amount of book learning.

THE WORKING-CLASS STRUGGLE FOR EDUCATION

So far I have treated education as an entirely bourgeois perquisite reluctantly doled out to the lower orders; but that is only half the story, and the old half at that. Three hundred years ago the bourgeoisie had demanded education as a road to power, now it was the turn of the industrial working class to do the same. The demand for education, the heroic efforts and the sacrifices made to get it, were an integral part of the labour movement, and, in the nineteenth century, were linked closely with the co-operative and trade union movements. 'Educate, agitate, organize' was the slogan of the Social Democratic Federation (p. 1095). All the great figures in the fight for Socialism were highly educated and very often self-educated men and women.

By the twentieth century this great drive had been slightly blunted by the concession of universal elementary education, and limited provision for higher education of the workers, such as Birkbeck College (incorporated in London University in 1920),[7.20] Ruskin College which, though at Oxford since 1899, has yet to rise to the dignity of a genuine

347. Workers' Educational Summer School at Balliol College, Oxford, 1912. A. L. Smith, the Senior Tutor and Dean of Balliol, spending a break with the students.

house,[7.148] and the Workers' Educational Association, formed in 1903.[7.158] The other danger was the almost irresistible impulse for a working man's child to use education as a way of getting out of his class, inevitably draining talent away from it in the best Platonic sense (p. 194). Nevertheless, the demand for more education still grew and gained force with the greater demands for skill and science in the new industries, and with the greater sense of working-class power. Towards the middle of the twentieth century the demand of the working class for access to all levels of education became irresistible.

The science of education, which had existed as an academic backwater for centuries, was now having to meet the requirement of educating the whole population. It must be admitted that it was not well constituted to do so. In part educational theory was a genuine attempt to find principles, largely psychological, behind the actual techniques of transferring information, and as such was no more and no less scientific than the rest of psychology. Even more, however, it was traditionally an attempt at a philosophy of education aiming at laying down its true purposes. As such it suffered from all the defects of the social sciences in an exaggerated form. Because it could not or would not recognize the changing character of society, or its class structure, educational theory, unconsciously as much as consciously, accepted that society as permanent, and aimed at finding ways of adapting the pupils to it. This inevitably made it conformist and apologetic.

INTELLIGENCE TESTS

An attempt to give education a more genuinely scientific air was meanwhile being made by the introduction of intelligence tests, originally taken over from some early efforts at scientific criminology. Because the quality of the performances of the subjects can be expressed in numbers, and these numbers can be added, divided, and subjected to statistical refinements, the results of such tests are deemed to be fully objective and scientific. The catch, however, is in the bias introduced in drawing up the tests themselves. In a class society any test applied to all children is bound to lead to class discrimination, and, as those drawing up the test must necessarily, as educated people, have had an upper-class bias, the results naturally stress the advantages of upper-class upbringing.[7.147] In any case, the greater contacts with parents with leisure, the availability of books, the possibility of visits and travel, give the middle-class child enormous initial educational advantages. If this head start, which operates quite independently of intelligence, is further confirmed by the results of intelligence tests, it must lead to an educational system grossly unfair to the working classes.

This is largely the way they have been used in the British post-war educational system so as to justify the continued segregation at the age of eleven of some 80 per cent of all children, a solid working-class group, deemed effectively incapable of higher education, while leaving to all those whose parents can afford to pay the freedom to proceed to the 'public' schools. What has been achieved as the result of this application of social science is a way of keeping the poor in their place and at the same time making them feel so inferior that they have not even the right to a grievance about it.

THE CRISIS OF SCIENTIFIC AND TECHNICAL EDUCATION

This admirable solution is now, however, becoming impracticable owing to the demands of modern technology on which economic survival depends. It was not designed, and naturally fails, to provide the numbers of scientifically and technically trained young men and women needed to run industry in the age of nuclear energy and automation. At last this deficiency is being realized with alarm in Britain and America, largely due to the impact of information of what has actually been achieved in this respect in the Soviet Union (pp. 1189). There has been a spate of official inquiries and reports, but very little has been done.[7.35; 7.71; 7.73] The country needs much more higher education and more scientific education at all levels. This implies more science teachers, hence higher pay for them and, in fairness, for other teachers as well, effectively doubling or more the budget for education. Moreover, the production of large numbers of scientifically trained people who know their importance in the economy threatens the domination of the humanistically educated upper classes. But to fail to carry through this educational revolution in time is to lose industrial and hence political pre-eminence.[7.128]

This has at last been realized, if only on paper, in that most backward educationally of the old industrialized nations – England. A blue-print for higher education in England, with emphasis on science and technology, has been prepared by the Robbins Committee.[7.70] It proposes to raise the number of young people in higher education from 180,000 to 560,000 in the years 1964 to 1980. At the end there would be a proportion in higher education equivalent to 17 per cent of the age group, a figure well below those projected in the Soviet Union and in the United States where it will be over 25 per cent in both cases by 1970.

13.5 The Ideological Background

PSYCHOLOGY

Of all the social sciences, psychology is the one which general opinion, particularly educated opinion, would concede has made the greatest advances in the twentieth century and has had the greatest influence in moulding the general attitudes towards life and society. Although the revolution in psychology associated with the name of Freud was a product of nineteenth-century thought (p. 1094), it was, until well after the First World War, but little known outside psychiatric circles. Its great influence on thought, and to a certain extent on human behaviour, came in only in the twenties. At that time Freudian psychology seemed a great revelation, as great as Darwinism in the last century, and, like it, a centre of fierce controversy, having ranged against it all the forces of respectability and religion. That controversy has now died down, but so has the belief in a new revelation on the working of the human soul. We are now in a position to estimate its place in human thought both in its genesis and in its consequences.

SIGMUND FREUD

The early work of Freud was in the natural sciences. He was a practising doctor, and had done research on the actions of drugs. When he took up the treatment of nervous cases he followed, in the first place, the methods of the experimentally minded French doctors, Charcot (1825-93)[7.63] and Janet (1859-1947). His material therefore lay in the real world, though in a very restricted part of it, for his patients were largely drawn from the very intellectual circles to which he himself belonged. But whatever the source of his material his ideas were part of the general positivist atmosphere of his time. Here, more than in any other branch of science, positivism influenced the whole interpretation and presentation of the results of his new technique of psychoanalysis

Though Freud himself was always striving to express his findings in terms of objective realities, the actual entities he constructed were far more like the spirits, virtues, and faculties of a medieval morality play than the material entities of the physicist and chemist. The unconscious, with its trinity of ego, super ego, and id, the complexes, the censor, the

348. Sigmund Freud; a photograph taken when he went on his first flight at the age of seventy.

libido, and the death-wish, were all invented by Freud to explain the strange imaginings, dreams, and compulsive behaviour of his patients. Just because these constructs had no possibility of material existence susceptible to any other form of demonstration, they tended to take on, especially with his followers, an absolute character, and created a fixed and almost mythological inferno of evil influences whose power to do harm could be mitigated, though never entirely removed, by the ministration of the analyst.

SOCIAL IMPLICATION OF PSYCHO-ANALYSIS

There is no scope within the limits of this book to discuss the principles of psycho-analysis, but only to note its social implications. These naturally took some years to reveal themselves. At first the new psychology met the most violent and hysterical opposition, emanating from the upholders of middle-class respectability and religiosity. It was only after the First World War, when the illusion of the eternal stability of a bourgeois existence had been shattered, that the influence of Freud began to spread so rapidly as to become, in a more or less diluted form, almost the religion of the intellectuals.

The reasons for its success must be sought in the objective insecurity of the times. Because it stressed the hidden importance of the irrational and primitive aspects of the human mind, the new psychology provided an explanation, indeed practically an excuse, for man's inability to deal with the problems of society. The upshot of Freudian psychology was that man was effectively ruled by his own unconscious instincts, implanted before birth and carefully nurtured for evil in the early upbringing of children.

It is true that the new theories of psycho-analysis had, and were intended to have, a considerable liberating effect from older dogma of the same kind, such as that of original sin. It was, at least in its early days, definitely anti-religious, and it provided strong support for a kinder and freer attitude towards children and sex. Nevertheless, the general effect of psycho-analysis on the elements of society that took it up with such enthusiasm after the end of the First World War was to distract people from attempting to solve social problems by any kind of corporate or political action and to draw them back into interests of their own personality, and particularly of their own sex lives.

Freud's own attitude was, and remained, essentially a scientific one; that is, his main concern was to find the simplest hypothesis that would guide him in interpreting the reactions of his patients and the reasons why they were relieved by his treatment. But the science he used was subjective and positivist and could only serve to multiply entities without reason. Towards the end of his life, when he attempted to extend his clinical ideas to the fields of anthropology and religion, his explanations were frankly mythical.[7.61] His followers, particularly those of them who broke with his somewhat rigid formulations, were to reveal far more clearly the mystical tendencies underlying the new psychology. With Jung it in fact returned not only to myth in the social sense, but to the concept of inherited myth and 'higher' truth of an occult kind, concepts which, in this or other forms, have been the basis of most of the Fascist movements of the twentieth century. Similarly, Adler, with his insistence on the power complex, was an unwitting prophet in the service of the enemies of his people. Now, half-way through the twentieth century, psycho-analysis has become almost respectable and is even making its peace with the Churches, dropping its overt insistence on sex to placate the parsons, but making up for it with a new infusion of anti-communism.

There is little need to say anything about other trends in psychology, for, apart from the neurological experimental psychology discussed in Chapter 11, it is either a rehash of the old Greek faculty psychology, or a more or less diluted Freudianism usually with a strong dash of mysticism. The role of psychology in the capitalist world is, as the examples of industrial psychology already quoted (pp. 1146 f.) show, one of providing a scientific excuse for the economic and political set-up. It also serves to discourage and disparage as emotional maladjustment any attempt to change it. As in earlier cases with the yogis and mystics (p. 179), the search for inner truth is a well-worn way to passive acceptance of outward evil.

PHILOSOPHY AND THEOLOGY

Any account of the development of the social sciences in the capitalist world in the twentieth century would be incomplete without some reference to the fields of *philosophy* and religion, or more particularly to the theoretical basis of religion – *theology* – known in the Middle Ages as the queen of sciences. This is not to say that one or the other are sciences comparable to those already discussed; they claim to be much more; they contain much less that is in any sense verifiable. The reason for including them here is that social science in the period of capitalism has still not fully emerged from the ideas and formulations of these pre-scientific forms of thought and feeling.

The theories or theologies of the religions of the world, especially of the Christian religion, have changed many times in the past (pp. 254, 292 f.) in response to the changing states of society. They may change again or fade away altogether in the future. What is impossible is to maintain an attitude to the world widely out of harmony with existing conditions: to try to bind the present to the past no matter in whose interest. This is the very definition of reaction and its intellectual counterpart, obscurantism. In so far as theology and philosophy attempt to do this they condemn themselves to sterility.

Although in the past philosophy and theology have figured as grand antagonists, like science and religion, in the struggle for the human mind (p. 662), they can now fitly be taken together. For, linked as they both are, in capitalist countries, to the interest of maintaining the *status quo* in a dangerous situation, they have composed most of their quarrels in the interest of common defence against the new materialistic philosophy. That this, however, can no longer be done by a rigid rejection of science and of economic and social trends has been shown by the recent developments at the centre of Western Christendom in Rome. At the Ecumenical Council called by Pope John XXIII (1881–1963) in 1962, and more clearly in his Encyclical, *Pacem in terris*, considerable concessions were made to the spirit of the times. He was particularly definite on the subject of the wickedness and folly of nuclear weapons, the sacred weapon of the Cold War:

Justice, then, right reason and humanity urgently demand that the arms race should cease; that the stockpiles which exist in various countries should be reduced equally and simultaneously by the parties concerned; that nuclear weapons should be banned; and that a general agreement should eventually be reached about progressive disarmament and an effective method of control.

Men are becoming more and more convinced that disputes which arise between States should not be resolved by recourse to arms, but rather by negotiation. It is true that on historical grounds this conviction is based chiefly on the terrible destructive force of modern arms; and it is nourished by the horror aroused in the mind by the very thought of the cruel destruction and the immense suffering which the use of those armaments would bring to the human family; and for this reason it is hardly possible to imagine that in the atomic era war could be used as an instrument of justice.

How far this will be accepted in practice by his followers and by governments that have a substantial Catholic element, such as those of France, Western Germany and the United States, remains to be seen.

It is a symptom of the general regression of the intellectual level that obscurantism and mystification in religion and philosophy are, on the surface, far more prevalent in the 'Free' world than they were fifty years ago. We are witnessing in our time a repetition of the change, from fashionable infidelity to fashionable religion, brought about by the fright the French Revolution gave to ruling classes. The step back now goes much farther and is more hysterical, because the present ruling class and their hangers-on are more thoroughly frightened than were those of 150 years ago. It is, for all that, even more superficial and tinged with hypocrisy, partly because a far more developed natural science has had all that time to penetrate into the general consciousness; but even more because there is now a living and growing alternative to the pessimistic obscurantism of the capitalist world that can no longer be ignored.

The intellectual retreat of the twentieth century was by no means exclusively marked by a return to religion, already discussed (pp. 1130 ff.). Especially in the early years after the First World War there was what appeared to be a rebirth of rationalism, in the form of flourishing philosophic schools, in the logical positivism of Russell and the Vienna School of Wittgenstein and Carnap, in the organicism of Whitehead (1861–1947) and in the pragmatism and behaviourism of Dewey and Watson in the United States. Nevertheless, in spite of a few gestures of defiance, the new positivism offered no serious opposition to the revived obscurantist drive; many of its exponents have even joined it. The enemy had changed; it was now no longer the Church or idealist philosophy, but the active materialism of the Soviet Union. By their logical criticism of the basis of all truth, the positivists did far more to shake men's belief in science than in their religious faiths.

THE PARTISANSHIP OF PHILOSOPHIC NEUTRALITY

In the past and particularly in the eighteenth century, religion had been attacked because of the absurdity of its beliefs, and had survived. A far deeper criticism was first fully developed by Marx, which has by now penetrated even into academic circles and touches the fundamental nature of religion itself, namely that of its social origins.[7.108] It is now becoming widely recognized that the attitudes towards the universe at large, and man's place in it, that have found expression traditionally in theology and philosophy are neither the results of abstract thought nor of divine revelation, but merely reflect the cumulative effect of a social tradition. Theology and philosophy have been built by human society in its own image.

Such an interpretation has always been vigorously repudiated by the official guardians of faith and learning, ostensibly in the name of a higher knowledge that can be completely detached from social consideration and that is given by revelation, intuition, or pure reason.

and is therefore objective and absolute. The maintenance of this position, harking back as it does to the dawn of conscious thought and ignoring the social advance of centuries of science, is just another aspect of the general mystification of the intellectuals in a decaying society. The ostensible reason for rejecting a social interpretation of philosophy and religion is not the real one. The underlying reason against admitting it is that it would open the present social system to attack.

Whether the arguments are based on intuition, faith, revelation, or pure reason has now become largely immaterial. What they have in common is more important than what separates them. It is at the root of the assertion of the existence of external and immaterial *beings* or *ideas* existing independently of society and not subject to change by human action. It is possible to make a variety of pretty patterns and systems of such ideas, and they can be sold either by appeal to the venerable traditions and wisdom of the Ancients, as is done by the proponents of Platonic or Thomist philosophy, or as the very latest discoveries of scientists and mathematicians who see through the deceptive appearances of vulgar matter, as is done by the professors of a variety of Neo-positivist and Logical Positivist schools. Their common objective, usually not clearly recognized and very rarely stated, is very similar to that of their prototypes in Greece, India, China, or medieval Christendom. It is to preserve the freedom and privileges of the cultured citizen, the brahmin or the churchman, and almost implicitly also to preserve the system of society that maintains them. As of old, those in power are usually willing to pay; for it costs very little for the moral and intellectual protection that philosophers and theologians provide for them.

THE WEAKNESS OF INTELLECTUAL REACTION

But the apparent strength of present-day philosophic and religious beliefs conceals a fatal weakness. Just because they are admirably adapted to justifying things as they are, they have lost their capacity for initiating change. Neither the sciences of Nature nor those of society can make use of them to bring about new advances. It has already been shown (p. 1093) that the immense labours of mathematical logicians and positivists have been barren of results in the natural sciences. The great advances that have been made in this century were achieved by experiment, explicitly or implicitly materialist, and closely linked with fully material techniques (pp. 746 f.).

In the social field the message of positivist philosophy has been even more conspicuously negative. By adherence to the laws of verbal and symbolic logic their newer protagonists have been able to demonstrate that statements which are neither analytic, like the tautologies of mathematics, nor empiric, verifiable by sense impressions, must be meaningless.[7.4] This refuses meaning to everything pertaining to social science, and denies the sense of religious, moral, or aesthetic values. By refusing the rank of meaning to most of what used to pass as philosophy, ethics, or aesthetics, the positivists did not, for the most part, aim at discrediting these disciplines. Their aim was rather to reduce their own field of discourse to logic and simple experience or later to logic alone. They left the rest to the operations of faith or mystical intuition, since they had proved to their own satisfaction that reason was powerless in these realms. The last section of Wittgenstein's *Tractatus Logico-Philosophicus* is entitled 'The Mystical' and ends with the words: 'Whereof one cannot speak, thereof one must be silent.' On such curious terms an unholy alliance between faith and reason has now been established.

For all practical purposes there is no quarrel between the positivist and the theologian in the social and political world. Both seek to prove the impossibility of a rational and historical account of society. Both seek to put limits on the range of human knowledge and achievement. In the place of scientific knowledge of a material world this would substitute an unacknowledged or acknowledged mysticism. Both are, in the most literal sense, *obscurantist*. Such attitudes of mind can only help reaction, however much their protagonists may hold themselves to be progressive and advanced thinkers. They tend to sap men's confidence to understand and master their own society and through science to mould the outside world to serve human needs.

It is here that the failure of the leaders of thought of 'Western Civilization' is most complete. They have no new solutions to offer to the great problems of the age: economic insecurity, colonial exploitation, and war. The old solution amounts to little else today than the indefinite continuance of capitalism armed to the teeth and wielding full police powers. No matter with what bright names this has been labelled – Freedom, Democracy, the Christian Heritage, the Open Society[7.131] – it conspicuously fails to inspire the mass of men and women with any devotion.

HYPOCRISY AND EVASION

Hypocrisy, conscious or unconscious, has never flourished as it does today. What is the use of talking of individual liberty in a society where almost every channel of expression is held by a few rich men or their subservient governments, and where success, livelihood, or even liberty, depends on having the right thoughts? What is the use of talking of equality of opportunity in a world in which more than half the people are deprived of the barest necessities of life, where they are hungry, diseased, and kept in ignorance, and in which most of the rest have dull and limited lives? What is the use of talking of ethics and charity or the sacredness of human life in a system based on the exploitation of such people for the benefit of a few, and where a major source of profit is the preparation of highly scientific means of blowing them to pieces, burning them to death, or poisoning them?

No wonder therefore that, to avoid the impact of such unpleasantness in the present world, the mind should be led to the contemplation of higher things or, failing this, allowed to regress to every old or new form of mystical nonsense. Even the forgotten astrology of the Dark Ages can now be made to pay a dividend. If anyone still cares to look

349. Astrology, clairvoyance and other forms of mystical superstition are still prevalent: a selection of advertisements from a currently published periodical.

they will see how far the present level of general intellectual appreciation in the capitalist world has fallen behind that of a hundred or even of fifty years ago. It is far less rational, sane, or hopeful.

PHILOSOPHY IN A WORLD AT PEACE

Hope in the future and faith in men's capacity to achieve it by their own efforts were preserved in the capitalist world only by the increasing band of those who had emancipated themselves from the limitations of this decadent and pessimist philosophy. Many of these had grounded their hope in the working-class movement and had participated in the great struggles, fiercer but more often crowned with success than ever before. Some had learned by that experience the meaning of the ideas and perspectives that Marx and Engels had worked out long before. And they are no longer alone. Many more who do not share their philosophy and who disapprove of much of their politics are finding themselves moved to protest against what has been done and is being prepared in their name.

Racial discrimination, colonial oppression, the waste of resources on armaments, and, worst of all, the prospect of general annihilation by the hydrogen bomb, have stirred even former supporters of the Cold War into protest. Many had given it their support simply because the alternative held up to them of totalitarian communism seemed even worse. But with the hydrogen bomb a point came when even this objection was transcended. Bertrand Russell had been one of those who had felt that peace was best secured through the definitive victory of the United States armed with atom bombs. In 1954, however, he realized the deadly potentialities of the hydrogen bomb and felt certain that it would be used in any future world war. Convinced that total annihilation would be an even worse alternative for humanity than communism, he urged the abandonment of war by all nations and the setting up of a world State. These views he expressed in his joint letter with Einstein and at the same time called for a meeting of scientists drawn from both sides of the Iron Curtain to discuss the dangers of nuclear war and the means for averting them. For him this was a most courageous decision to take, and the response showed that he was expressing opinions shared by many others in the world of science and outside it.

PUGWASH

The Russell–Einstein letter was the starting point of the first serious move to get the scientists of the world together in order to soften the

350. Participants to the first Pugwash Conference. They are (left to right): I. Ogawa, Chou Pei-Yuan, V. P. Pavlichenko, S. Tomonga, C. F. Powell, A. M. B. Lacassagne, A. V. Topchiev, A. M. Kuzin, E. Rabinowitch, G. Brock Chisholm, D. V. Skobeltzyn, J. S. Fraser, C. S. Eaton, H. J. Muller, J. Rotblat, H. Thirring, L. Szilard, W. Sekove, E. H. S. Burhop, M. E. Oliphant, M. Danysz.

impact of the Cold War and to look for terms for possible co-existence. Although at first it gave rise to much suspicion, in the end it was found possible to bring together not only progressive scientists from the academic establishments but even those engaged in defence science. Thus serious, though strictly unofficial, discussions could be held on such difficult problems as disarmament and nuclear tests, and also on positive alternatives for using science for relieving poverty in the world. There have been about a dozen such meetings, and there is no doubt that the attitudes of governments on both sides of the Cold War have been seriously modified by their findings. What Pugwash has done has been to bring the reality of nuclear war much closer to the consciousness of millions of people and at the same time to provide an alternative that is at least practically workable:

We believe it to be a responsibility of scientists in all countries to contribute to the education of the peoples by spreading among them a wide understanding of the dangers and potentialities offered by the unprecedented growth of science. We appeal to our colleagues everywhere to contribute to this effort, both through enlightenment of adult populations, and through education of the coming generations. In particular, education should stress improvement of all forms of human relations and should eliminate any glorification of war and violence.

Scientists are, because of their special knowledge, well equipped for early awareness of the danger and the promise arising from scientific discoveries.

Hence, they have a special competence and a special responsibility in relation to the most pressing problems of our times . . .

. . . The increasing material support which science now enjoys in many countries is mainly due to its importance, direct or indirect, to the military strength of the nation and to its degree of success in the arms race. This diverts science from its true purpose, which is to increase human knowledge, and to promote man's mastery over the forces of nature for the benefit of all.

We deplore the conditions which lead to this situation, and appeal to all peoples and their governments to establish conditions of lasting and stable peace.

From the Statement of the Third Pugwash Conference, Kitzbühel, 1958

The movement continues and spreads and it should be clear by now that if it depended only on the judgements of intellectuals, the breach of the Cold War would be relatively rapidly healed. However, it would be thoroughly unrealistic to expect that this would be sufficient to overcome the enormous powers of economic vested interests and the psychology they have created in the peoples of the world.

It is indeed now becoming, quite apart from the suicidal nature of nuclear warfare, difficult to justify the sharp division of the world into two necessarily hostile régimes. The new Soviet leaders have admitted the existence of long-suspected injustices and oppression, and in ending them have already taken serious steps to return to fuller freedom and democracy and to find a guarantee against any return to personal rule. The old black and white distinctions of the Cold War, whichever side it was viewed from, are not so sharp as they were. It has effectively now lost its ostensible *raison d'être*. Other questions, notably those of Asia, Africa and Latin America, are cutting across old divisions. The peaceful influence of India is becoming more and more impressive. Intellectuals are beginning to grasp some of the implications of a world without war, particularly, as Sartre was the first to point out,[7.13.263] that the abandonment of the Cold War logically demands not moral isolation but active co-operation and understanding.

13.6 The Social Sciences in the Socialist World

That understanding must extend in the first place to what has happened in the building up of the first socialist state, the Soviet Union, a vast social experiment the influence of which has since been reflected in the other socialist States, and far beyond them. This opens a new field to social science, for, as we have seen, the social sciences of the past grew up and inevitably had to deal with class-divided societies. Here new social forms are evolving, far more consciously designed than the traditional developments of older societies. We are witnessing the birth of a social science that can be experimental as well as analytical.

To understand the nature of the new socialist societies that came into being in the aftermath of the First World War in the Soviet Union, and later of those that grew after the Second World War in many parts of the world, it is essential to say something about the genesis of the Russian Revolution itself and about the man who more than anyone else was responsible for the shape that it took and for its influence in the world of the twentieth century, Vladimir Ilyich Ulyanov (1870–1924), known to posterity as Lenin.

THE GENESIS OF THE RUSSIAN REVOLUTION

Marx and Engels had been the formulators of the criticism of capitalism and the indicators of the path of revolution which was to lead to its supersession by a working-class state. They were not, however, to see the fulfilment of their prophecy. That was to be left to the first breakdown of capitalism which occurred, not in the old, industrial capitalist countries, but in the relatively backward area of the Russian Empire.

Russia, at the end of the nineteenth century, was politically and economically at the same time an imperialist and a colonial country. With an old culture and tradition, it was politically the most backward State in Europe, and the Tsarist autocracy was occupied in a policy of aggrandizement in Central and Eastern Asia no less imperialist than that being pursued by Britain in India or by all the European powers in Africa. But this imperialism lacked its essential economic base; Russia was industrially backward, and though, as time has shown, quite capable of developing its own resources, this was impossible under Tsarist government. Instead it had become a chosen field of investment by which foreign firms exploited its resources for their own benefit and multiplied loans to the Government on exorbitant terms. It is not surprising that in the resulting atmosphere of misery, graft, and frustration almost everyone of intelligence and spirit was in opposition to the system, which could only be maintained by a brutal and stupid police repression. Most of this opposition, however, lacking an adequate social theory, spent itself uselessly in conspiratorial terrorism. A turn came when, simultaneously with the growth of an industrial working class, Marxist ideas began to penetrate into Russia. Among those who took them up most intelligently was Plekhanov (1856-1918).

LENIN

His most important disciple, who far exceeded him in intelligence and ability, was Lenin. He came on Marxism as a student and already a revolutionary. It was he who, by first absorbing basic Marxist theory, found how to apply it not only to Russian conditions but also to the economic, political, and intellectual world situation of the turn of the century, already very different from those Marx himself had dealt with. If he had not been one of the greatest of the world's political leaders, Lenin's intellectual pre-eminence would have been recognized by reason of his contributions to economics and philosophy. He was the first to see in the rampant *imperialism* of his time, which even the progressive leaders of Fabianism supported (p. 1099), the symptom of the decay of capitalism unable to find sufficiently profitable home markets for its capital goods production. He was also very much alive to the intellectual and cultural changes that accompanied the era of imperialism. In his *Materialism and Empiro-Criticism*[7, 95] he analysed the positivist tendencies of Mach and his followers in the Russian socialist movement, and showed that, for all their claims to an advanced and objective scientific outlook, they were on a path that could only lead back to the pure idealism of Berkeley and Plato, and hence to the support of the position of reaction.

A PARTY OF A NEW KIND

For Lenin, however, the central problem was the political one. How was the revolution, which all Russian progressives agreed was necessary, to be carried out, and what was to follow it? Lenin had learned from Plekhanov the importance, even in backward Russia, of the industrial workers as a revolutionary force. He drew from it the practical conclusion of the need for a *party* imbued with a Marxist understanding of the tasks before it and devoted to winning the people for the complete triumph of socialism. This party was originally the Social Democratic Majority (Bolshevik) Party, later to become the Communist Party. It met its initial trials in the revolution of 1905, an aftermath of the disastrous Russo–Japanese war. This revolution was in the first place a bourgeois one, though in order to shake the Tsarist Government the bourgeoisie made common cause with the workers. It failed because the bourgeoisie preferred to compromise with Tsarism rather than to push the revolution further.[7.96] Lenin showed his real greatness after that defeat, which did not undermine his confidence in ultimate success but which taught him many valuable lessons, particularly on the need to isolate the bourgeoisie by making an alliance between workers and peasants. The turning point came when he was able to transform a second revolution, which was in its first stages also a bourgeois revolt brought about by the military failure of Tsarism in the First World War, into an effective and permanent seizure of power by the workers.

From this, despite all the efforts of external and internal enemies, he was able to bring into existence for the first time in history a new kind of State with the people, led by the workers and peasants, in control. As both the theorist and the architect of the new society Lenin holds a unique place in human history. Because he was able to use the test of practice he was able to check and correct his earlier views. He continued his political analysis even in the midst of the struggles of the early days of the Soviet State. In *The State and Revolution*,[7.97] published on the very verge of the upheaval, he had analysed the difference between the bourgeois State as the executive organ of finance capital to which the people could at best be persuaded to give their passive assent through the elaborate forms and fictions of parliamentary democracy, and the new socialist State, which must draw its strength from the active participation of all the people in thousands of organizations, factories, streets, and farms. This new form of democracy would, in the first place, break with the whole State and economic machine of former governments, tied as it inevitably was to the interests of capital. But a new State had to take its place with greatly extended powers, directing in a planned way the economic functions of production and distribution. Further, as long as it was battling for existence in a hostile capitalist world, it had also to build up and maintain the military and diplomatic functions of government.[7.97]

Lenin also saw very clearly the place of the Russian revolution in relation to the world socialist movement. He saw that in Germany, where after its defeat the conditions for revolution were most favourable, the attachment of most social democratic leaders to capitalism was certain to betray the country to the forces of reaction.[7.98] He realized that when interventionary wars had failed to break the new republic there would follow a long period of coexistence, in which it would be necessary to deal with capitalists on a strictly business basis, while depending on the friendship of the peoples of the capitalist countries to prevent further attacks.

351. A banner with Lenin's portrait being carried in a procession to Red Square.

This he knew would be no easy task. The barrage of hysterical abuse against the new régime by the controlled Press of the capitalist world never ceased, and its effect was being reinforced by the activities, sometimes merely stupidly enthusiastic, sometimes deliberately provocative and wrecking, of some of the friends of the revolution abroad. In his

Left-Wing Communism, an Infantile Disorder, Lenin analysed and denounced these tendencies, and showed the absolute necessity of a sound social theory in order to avoid on the one side the over-caution and on the other the rashness which, when they were honest, flowed both from a lack of understanding of the social forces and from the movements they generated. His warnings were to be amply borne out by the political events inside and outside the Soviet Union in the subsequent thirty years. The writings of Lenin are still a major source of the new social science of the present and the future. They have, however, been liable at different times to different interpretations, such as those at present in dispute between the Soviet Union and the People's Republic of China. Lenin, himself, had always proved in practice to be a man capable of determination and also of compromise, according to the circumstances. What he would have done in the present circumstances is naturally impossible to predict in any scientific way. We may be reasonably sure, however, that he would have found the way to unite the differing factions round his life's task of establishing and defending socialism. His work remains an inspiration and an example to thousands of millions of people in the world today.

The treatment in this section of social science in the socialist countries must necessarily differ from that already given for the capitalist countries. This is because we are here dealing with societies in rapid transformation and evolution and, what is more, in evolution along lines which follow the ideas sketched out in the nineteenth century by Marx and in the twentieth by Lenin.

While the philosophers, the economists, the historians, and other social scientists in capitalist countries were caught in the maze of disturbing events that preceded and followed the Second World War, their colleagues in the socialist world were occupied with very different tasks, facing a multitude of new practical problems. They were concerned with material and social creation, with the building up, despite the most discouraging difficulties, of a new kind of life, in the factories where workers had a positive and immediate interest in better production, in the collective farms, in the new uprising of native enterprise in Asia, in the great social services of health and education. They also had to live through the strains of an emerging society beset by external enemies and troubled by internal struggles.

THE FIRST SOCIALIST STATE

For the first thirty-four years after the Revolution, the Soviet Union was the only socialist state in the world. Its peoples had, from their own

352. The Russia of the Tsars had been industrially backward, but new industries were started and others expanded after the revolution. Blast furnaces at Stalinsk in Asiatic Siberia, north-east of Kazakhstan.

resources, to build a new kind of society and to find out how to do so as they went along. Its whole history has been one of enormous constructive effort against persistent internal difficulties, natural and social, and against the bitter enmity of the capitalists, as exemplified by wars, economic blockades and unceasing hostile propaganda and attempts at subversion. To be able to hold out at all and to build a new civilization in the face of difficulties required a degree of determination, discipline, and endurance which no people have ever had to show before.

Naturally the great events of those times coloured the attitudes and doctrines in the social sciences to an even greater degree than did the contemporary events in the capitalist world. For here economic and political thought was no academic ornament but a vital part of daily life and one of the great tasks of socialist construction. Events, indeed, often moved so fast in the Soviet Union that it was difficult for reflective thought to keep up with them and it is only in our time that it will be possible to analyse the formative effects on social and economic thinking of the actual changes that took place between 1917 and the present time. I am clearly not the person to carry out such an appreciation, lacking both the time and the access to the original sources. All I can do is to indicate very briefly what I consider were the chief influences and to leave to others the task of evaluating their effects in detail.

THE REVOLUTION OF 1917 AND ITS EFFECTS

The greatest and most persistent influence of all was that of the Revolution itself, carried out by the Communist Party, which Lenin did so much to create. Lenin's thought and active direction coloured everything that was to come after. Although at times his message was obscured and to a certain extent narrowed into dogmatism, it could never be lost. It was not even so much in his teaching, but in his expression of it in practice that Lenin provided his most precious legacy. However difficult and dangerous the situation, whatever desperate measures had to be taken to deal with it, Lenin had always trusted the people, had listened to them and taken them into his confidence, so that he carried with him the understanding and the will, as well as the hope of his people. It was the forgetting of this lesson that was to be the source of some of the worst features of Stalin's personal rule.

Stalin (1879-1953) himself had, apart from his historic services in the period of the transition to socialism, a considerable role as a formulator of socialist theory: in particular, his formative ideas on nationality and his insistence on the need for a powerful economic base for the subsequent transition to communism. The excesses of his personal rule, his suppression of criticism and his arbitrary persecutions, however, did much to weaken the original revolutionary impetus of the Soviet people but did not, as the Second World War showed, by any means extinguish it.

We tend to see history too much in terms of a few individuals, partly because it is only when condensed in this form that it can properly be grasped by the mind. The real influence of the Revolution is the compound of innumerable acts of the struggle, creation and devotion to a cause, of millions of individual men and women gathered together in their organizations – the Party, the trade unions, the collectives. To the Party fell the heaviest material and political tasks of restoring the economy of the country and of repelling foreign intervention. It had, moreover, to do so without the help, and often with the open or secret hostility of former administrators. It had to discover and use without delay new social forms fitted for a society that had never existed in the world before. The Party members could never have accomplished what they did if they had not already possessed a key to the understanding of social dynamics, a way of liberating the initiative and full capacity of millions of human beings in co-ordinated action. This they found in the works of Marx and Engels, which they came to understand and to extend on the basis of their own experience. It was the Communist Party that had effectively to carry out this momentous step in human history – the creation, the preservation and the expansion of the first socialist state in the world.

THE SOVIET UNION – A MULTI-NATIONAL STATE

The Soviet Union itself became the first example of a multi-national State, liberating historic nations and creating others from what had been previously only oppressed peoples and tribes. This example, which has now been followed by China and will be in many other areas of the world, opens the way to a reconciliation between the cultural and historical individuality of national groups and their political and economic interdependence.

353. The geographical extent of the USSR is large, covering some 8·7 million square miles, an area almost three times that of the United States and over twice that of the People's Republic of China. The ethnological types embraced in so vast an area vary widely, as this photograph taken at the Kazakh Institute of Mining and Metallurgy indicates.

The nationalities policy was, in time, to have a notable effect outside the Union. It provided a most eloquent example to the suppressed nationalities of colonial and imperialist dominated countries throughout the world. They could see how it was possible in practice to assert at the same time their right to their own cultures and to advancement, through their own efforts, into the ranks of rapidly advancing modern industrial nations. This is the secret of the new confidence of the movements of national liberation in Asia, Africa and Latin America. Later still, it has provided the inspiration for the Bandung union of Asian and African countries who, as part of the great 'Neutral Bloc', have so changed the face of world politics and broken up the sharp division of the Cold War.

PLANNED INDUSTRY AND COLLECTIVIZED AGRICULTURE

It was in the planning of industry, particularly of heavy industry in 1928, that the Soviet Union made its greatest contribution to the development of material culture. Henceforth socialist principles were to be embodied concretely in practice and were to demonstrate, despite all obstacles, the superiority of the socialist objective of planned production for use, over competitive production for profit, which was by no means improved when this competition had been modified by the creation of monopolies.[7.13.143] The discovery of the most efficient methods of planning, involving such problems as the utilization of natural resources, the recruiting and training of a highly skilled labour force, and the proper balance between central and local organization, had largely to be worked out in practice and thus to form the basis of a new kind of economy. Despite many mistakes the lesson has been learned, as witness the speed and steadiness with which socialist is overtaking capitalist production. And this lesson is a general one; it is rapidly applicable to any under-developed country which is run on socialist principles, as the rapid advance of industry in China shows. Actually, it was easier to apply there than in the apparently more advanced countries of Eastern Europe, in which the older capitalist modes were more entrenched and where the mistake was made of trying to build massive heavy industry in countries too small to have sufficient natural resources and manpower. These mistakes are now being overcome through economic co-operation, and the superiority of the planned approach is again being vindicated. It has already had a profound influence not only on under-developed countries such as India and the new states of Africa, but also on the industrial capitalist countries.

Closely following the beginning of industrial planning in the Soviet Union, and a necessary complement to it, was the reform of agricultural methods in the setting up of collective farms in the place of the isolated and unequal peasant holdings set up in 1917 after the break-up of the estates of the nobility. Without such collectivization, it would not have been possible to use the, at first scanty, agricultural machinery to provide enough food for a population more and more engaged in industry. The method of the collective farm is not the only way of doing this. It has itself been much modified in the Soviet Union now that more resources are available.

In China, on the other hand, the institution of communes aimed at the combining of local industry with agriculture, thus solving the perennial problem of oriental agriculture, rural under-employment, still the curse of India. In their successes as well as their failures, in the face of unprecedented natural disasters, to maintain the increase of agricultural production, the Chinese communes have valuable lessons to teach. There is no way out for the hundreds of millions of primitive agriculturalists under feudal domination or on foreign plantations, except by the abolition of these conditions and by the setting up of

354. Peasants' new housing in a suburban commune at Hangchow, east China. At such communes the aim is to combine local industry with agriculture.

some form of socialist co-operation. Yet this, by itself, will not be able to relieve the peasants from unending toil or to free them from the violent vagaries of climate, without machinery and fertilizers, that is, without access to centres of heavy industry.

THE WAR: ITS SHADOW AND ITS AFTERMATH

The attitude of the peoples of the Soviet Union towards the outside world, as well as that of the outside world to them, would have been very different if Soviet economic and social development had been allowed to proceed in peace. Actually, the Union was never free from passive or active interference from capitalist countries, culminating in the Second World War which devastated the richest of its territories and caused the deaths of some thirty million soldiers and citizens. During the same period the people of China suffered from the oppression of their own landlords and bourgeoisie, aided and abetted by capitalist powers, and from almost continuous warfare, both civil and against the Japanese imperialists.

These events go far to explain, if not fully to justify, the internal policies of the Soviet Union during the years of strain that started with the advent of Hitler to power in 1933 and merged in 1950 into the new strain of the Cold War, with the United States as the chief antagonist. In a very real sense, the peoples of the Soviet Union during all this period have been in a state of siege which, though it may be eased, only the ending of the Cold War under the influence of world popular opinion can definitely lift.

Only a government imbued with the greatest determination could hold out in these conditions, and this goes far to explain the acceptance of the personal rule of Stalin, and his glorification, despite his many unjust and arbitrary actions. With the increasing successes of Soviet economy, that phase is now ending and we are seeing already the fruits of education and science in the easier and more democratic direction of N. Khrushchev. The similar difficulties that have beset the young Chinese People's Republic, beginning with the Korean War, have also called for a discipline that is essential to establish a socialist State that can be viable in a world still largely dominated by capitalism.

These considerations must be taken into account in any appreciation of the quality of social thought in a socialist civilization which is still in a condition of fighting for its existence. No one is yet in a position to present a justly balanced picture, weighing the difficulties of the objective situations against the errors, distortions of principles, and actual injustices of socialist governments in their attempts to cope with them. To do so, however, would be an undertaking of the greatest importance, as it touches the very essence of Marxism and the particular problems – which Marx and even Lenin could only dimly foresee – of government during the building of socialism and its transition to communism. What is needed is not praise or denunciation, but deeper understanding of the relations of personalities and movements in a rapidly changing society set in a war-shaken world. This is one of the major problems of political science and, in the measure it is solved, will provide warnings and lessons for further action. No one who has the cause of human welfare at heart can fail to be saddened at the sufferings and injustices that have accompanied the building of the first socialist states. Some were inevitable; to avoid the others, the lessons of how they came about must be, and are being, learned. In any case their occurrence can give little real comfort to the enemies of socialism, for it is evident that the Soviet Union has survived, despite all its internal and external troubles, and that it has shown enough inner strength to set about reforming itself.

This reform is continuing and rapidly moving into a new phase corresponding to the possibilities given by the Seven Year Plan of advancing simultaneously in heavy industry, consumer goods and housing. The Soviet Union of the future, as well as all the socialist countries associated with it, will be fully capable, as no capitalist country can be, of making full use of the new scientific-technical revolution and all it implies in the transformation of human society and its laws.

TOWARDS A NEW SCIENCE OF SOCIETY

The fact that a break has been made with the past, and that such momentous changes are taking place and are still in their earlier stages in the Soviet Union, makes it difficult and unprofitable to discuss in any detail the state of the social sciences there. Nevertheless, the history of their development can be followed in broad outline, with particular stress on the new lines of social thought and analysis that have emerged.

The wide differences in the material and political circumstances of social scientists in the Soviet Union from those in capitalist countries (p. 1182) accounts for only one part of the differences between the pursuit of the sciences themselves in the two cultures. There are other historic reasons: under Tsarism the Marxists had naturally never played a large part in official and cultural circles. The advent to power of a government inspired by Marxism and backed largely by industrial

355 a, b. The Second World War brought devastation to the East as well as to the West and the heroic stand made at Volgograd (Stalingrad) has been echoed in the astounding rebuilding programme.

workers led naturally to an effort to transform the social sciences and to give them a Marxist base. But the process was slow, difficult, and liable to mistakes. From being the doctrine of a more or less persecuted minority, Marxism now enjoyed official approval. It would be wrong, however, to treat social sciences in Russia as representing merely the continuation of the revolutionary Marxist tradition. A long process of assimilation was first needed, all the slower because all the most able Marxists were, from the time of the Revolution onwards, largely employed in administrative and political tasks. What happened in the first stage was an absorption of Marxist ideas by already formed social scientists and, only after the first twenty years, the appearance of younger men and women trained largely under Marxist auspices. Loyalty and even enthusiasm for the building of the Soviet Union did not necessarily imply understanding or acceptance of Marxism.

The very subject of these sciences, human society and its culture and institutions, was radically changed by the Revolution and continued to evolve more or less rapidly in the ensuing years. This necessarily imposed a demand for new methods of study and opened the way to a more creative use of Marxist ideas. Scholars in the older disciplines of history, anthropology, and philology might, it is true, continue their labours along more or less familiar lines, though with a new inspiration. Economists, on the other hand, had to begin their studies anew with a radically different basis from the market and profit economies of the past, and that no longer as observers but as participants in the creation of new economic forms and measures of which the merest outlines were to be found in the Marxist classics. As to the relatively newer disciplines of sociologists, psychologists, and educationists, they never underwent the same academic development as in capitalist countries, and the work of their exponents was only one part of the practical business of building and educating a new society. Political science was even more thoroughly merged in actual politics. What took the place, to a certain extent, of these sciences was the beginning of Marxist philosophy (see pp. 1182 ff.). Here, however, as a somewhat paradoxical result of a government and party professing Marxism being in power, official Marxist philosophy became so overloaded with political responsibility that it showed far less originality and development than it had when Marxists were in opposition, It does indeed seem that less was done in the development of Marxist theory in the last twenty years in the Soviet Union than was done outside, especially in China, but also in Italy, France, Britain, and other European countries.

Taken in all, therefore, it is probable that there is more to be learned from the actual developments of society in the Soviet Union than from the study of the conclusions of academic social science there. Where everything has to be changed or made anew there are for every professional social scientist thousands of practical people who do not even know that they are social scientists who are occupied, in the course of their life and work, in remoulding and creating social forms and who are learning as best they can from their successes and mistakes what are the limits of possibility in dealings between human beings. While people are actually remaking their society the theories they work on, as far as they are explicit, are likely to be simple. The thorough scientific study of society must begin when the change has settled down a little and recognizable patterns appear. However, the risk of an incomplete treatment must be taken, for the development of social forms and social sciences in the Soviet Union and its bearing on the whole question of the social aspects of science is so important that it cannot be omitted.

Despite all these qualifications, it is becoming evident that there are emerging from the Soviet Union both new experiences and new sciences of society which are making their influence felt more or less directly throughout the world. As an example I might cite the concept of economic planning, which is not only the rule in the socialist world but has spread to many under-developed countries such as India, Egypt and Algeria, and has now been adopted in many of the older capitalist countries, notably England.

THE RE-EVALUATION OF HISTORY

Nor is this influence only concerned with new social forms; it involves a study of the past and of other current societies. One feature of this is in the renewed interest in history and the allied sciences where the use of Marxist ideas has produced not only new interpretation but brought to light new material. These studies have dealt, in the first place, with the cultural heritage of Russia and of the other nations of the Union. History, archaeology, and anthropology have flourished as never before. Soviet archaeologists have uncovered, by well-planned and well-executed expeditions, unknown regions of the past. Much of this work, though published, is not sufficiently known in the rest of the world. Already we know, however, that from the Neolithic period onwards unsuspected cultures and civilizations have flourished in Russia and Central Asia. The new knowledge, which goes far to explain the successive waves of folk migrations, should already be an integral part of world history.

356. Modern Soviet revaluation of history includes studies in anthropology and archaeology, as well as history. The photograph shows a wooden burial chamber in a barrow at Borshevo on the River Don. Photograph from *Archaeology in the USSR* (Penguin Books, 1961).

Even more relevant, because closer to our own problems, is the careful re-evaluation of the history of Europe, of which that of Russia, despite all the denials of the paladins of Western Christian Civilization, has always been a part. The relation of the early Russian tribes to the Northmen on the one hand and to Byzantium on the other has been traced and the distinctive national character established. But the work of the Russian historians has not stopped there. Professor Kosminsky's studies, for example, on medieval history in England and France are acknowledged by scholars to have thrown new light on the growth and decay of feudalism. The new methods of historical research based on Marxism are already bearing fruit outside as well as within the Soviet Union.

HISTORY OF SCIENCE IN RUSSIA

Of special interest, because so neglected in the past, has been the new interest focused on the history of science and technology. This started early in the Soviet Union, and the contribution its delegation made to the History of Science Congress in London in 1931 had a profound effect in revealing a new approach to science as a social and economic phenomenon, rather than as an expression of absolute and pure thought.[7.144a] It may be said that this impact has set going a whole new school which together with its critics has established the importance of the social history of science. Since then more profound studies have broadened the picture and removed early crudities. Particularly notable, for instance, is the work of S. I. Vavilov (1891–1951) on Newton (p. 473),[7.170] which gives a deeper insight into his work and a more intelligible account of the origin of his ideas on chemistry and atomic theory, which had hitherto been neglected.

Naturally, most of the interest has been focused on the history of science in Russia, a subject unjustifiably neglected by most earlier historians of science. Here the exaggerated claims for priority that had been made for Russian scientists as part of the post-war campaign for boosting national morale should not blind us to the value of the researches carried out. They should lead to a general recognition of the very real contribution of Russian scientists throughout the centuries. As well as famous names such as those of Lobachevsky, Mendeleev, and Pavlov, Russia has produced creative scientists and inventors such as the chemist Butlerov (1828–86), the physicists Lodygin (p. 613) and Popov (p. 774,) and the aircraft pioneer Zhukovsky (1847–1921), to cite no others. Great attention has also been paid to the history of technology, far more, indeed, than in Britain, the country of origin of the Industrial Revolution.

THE PLACE OF THE SOCIAL SCIENCES

Outside anthropology, which is well studied, sociology and social psychology have, as far as I know, not been much developed in the Soviet Union. Prejudice against their apparent capitalist character has to some extent prevented the development in the Soviet Union of social sciences of a new kind, based on a Marxist appreciation of the importance of historical and economic factors, but employing more honestly and intelligently than in capitalist countries the new techniques of statistics and factor analysis developed there (pp. 1139 f.). The task of doing so is all the more important in that, in a society that is rapidly growing and producing new forms of organization, there is the possibility of developing a dynamic sociology which, if it has any scientific value, should help to make the changes easier and to avoid mistakes.

Particularly important in this respect is the part that social psychology should take. It is no longer enough to say that remodelling productive methods and the corresponding productive relationships should by themselves remove the causes of psychological maladjustments. Indeed, as we know now, the methods of force and crude propaganda, even where used with the best intentions and for the best ends, intensify the troubles they are designed to cure. The present policy of return to reason and persuasion will make great demands on the development and practice of an effective social psychology. For this we have already had many practical examples in the building of new collectives and other working groups, as exemplified, for instance, in the life and work of Makarenko (1888–1939) (pp. 1188 f.). Men like him have shown that if every man or woman comes to experience and understand his importance to society and has the chance to make his opinion weigh in his everyday work, much of the inferiority and frustration which have characterized class societies up to our time will disappear.

SOVIET ECONOMICS

The science of economics has been the centre of Marxist thought. It was, however, even up to Lenin's time, almost entirely concerned with critical analyses of capitalist economics, and it has proved more helpful to the new society in showing them what to avoid than telling them what to do. The whole history of the Soviet Union has been one of economic innovation – nationalized industries, collective farms, investment and foreign trade in the hands of the State. At the same time commodity production from private holdings continued. The adaptation of a system of banking to finance these enterprises had also to be worked out. From this ultimately successful achievement much should be learned, if only to find the common irreducible elements of capitalist and socialist economics: for instance, a theory of costs and prices when the latter do not depend on a market but can be fixed, within limits, according to government policy. This naturally complicates the question of wages. In post-war years the government has preferred to raise real wages by lowering prices so as not to interfere with the high incentives given to skilled workers. The tendency now, however, seems to be to level up, by raising the wages of the lower-paid workers and giving the trade unions more responsibility for fixing wage rates.

Much has been written on economic questions in the Soviet Union, but I know of nowhere where it is available in a comprehensive form in translation. The official text-book on Political Economy is manifestly inadequate in this respect: it has been much criticized and recently has been sent back for complete revision. It seems, accordingly, that socialist economic theory is still incompletely worked out and liable to considerable controversy. One difficulty that has faced economists in the Soviet Union till recently is the unavailability of most economic statistics, which have been kept secret for security reasons and have only recently been released. Some years of the new régime will be needed before the lessons of a freer and more decentralized economy can be learned.

A PLANNED ECONOMY

The most important innovation in Soviet economics has been the practice of *planning*, at first annually and then for five-year periods. The plans are the expression of the basic socialist purpose which was formulated by Stalin just before his death,

. . . the securing of the maximum satisfaction of the constantly rising material and cultural requirements of the whole of society through the continuous expansion and perfection of socialist production on the basis of higher techniques.

To achieve this result the planning of production and consumption in some detail is necessary, in place of the profit motive and the free market, which serve to produce not only the booms but also the slumps

357. The huge hydro-electric generating station on the Volga which has an output of 2·5 thousand megawatts.

and wars of the capitalist world. Conscious calculation of the optimal distribution of productive resources, material and human, to social consumption over long periods has thus been substituted for the chance interaction of many factors. This advance is of fundamental scientific importance. It represents a higher stage in social evolution much as the appearance of a central nervous system did in organic evolution.

To reach this higher stage is in itself a difficult task. Only by trial and error has it been possible to build up over a number of years a body of experience which by now has almost made it a routine operation. The present post-war drive to achieve a higher standard of consumption without slipping back on the industrial front requires a combination of planning based on previous experience and a close check on the development itself. The word plan suggests to the enemies of planning, and to many more who have been deluded by them, that plans are something rigid and artificial imposed on society. The Soviet plans never were that; such a conception would run quite counter to the whole spirit of Soviet life. Each plan marks an intention which millions of people have, through their organizations, co-operated in defining, to carry out a common task for a common benefit. In doing so they meet difficulties largely unforeseen, but they also see unexpected opportunities. In both cases the plan is modified, and the people taking part in it are not just executing rigid instructions, but are perpetually ready to improvise and accelerate. We, with our orderly and slow methods, find it difficult to imagine how anything can be done in such a way, but in fact the plans have succeeded: the towns, factories, canals, dams, and railways have been built, and far quicker than any foreign experts thought was possible.

This speed of carrying out of the plans, particularly in the earliest phase of industrialization and collectivization, has given rise to criticism of another kind, based on the cost in social dislocation and personal suffering it has caused. It is now admitted that some of this might have been avoided, if more democratic and persuasive means had been employed (pp. 1174 f.). Nevertheless, a rapid change in itself could not be avoided, and the extreme speed was a necessity forced on the Soviet Union by outside forces.

The experience of other under-developed countries such as India, once they have been relieved of the economic stagnation that was impressed on them by imperialist exploitation, has shown that the minimum rate of investment in heavy industry is about 6 per cent. For such a country to invest at a low rate or, in other words, to leave most primitive methods of production as they were, in a world where other countries are advancing faster, is to ensure a rise in the relative cost of production. Competitive selling becomes difficult, and the country, becoming more and more backward, is driven into ever greater dependence and foreign exploitation, as was, for instance, China before 1949. To attempt a slow transformation in which the most conservative interest is respected is effectively to prevent the change altogether under guise of promoting it. This is especially true of changes of ownership in agriculture, where the power of the strong farmer or *kulak*, whose favour or ill will can make or ruin a poor man, can be shaken only by resolute and rapid common action of the poorer peasantry. Analogous difficulties have wrecked land reform schemes in such diverse countries as Italy and India.

For these reasons the Soviet agrarian revolution, and the industrial construction which was its indispensable base, could only have been put through in a very few years. Moreover, even if it were granted that a slower movement might have succeeded, there would have been no time for it to do so. Long before it could have been ready the Union would have been overcome by the Nazi attack, and the civilization of the whole world would have received a set-back for many decades.

SOCIAL RESPONSIBILITY

Economic planning requires for its success a degree of *social responsibility* never before reached. Its achievement poses problems both for governments and for peoples. Administrators are always under pressure to take short cuts and to compel when they cannot persuade. People are under temptation to seek private advantages, which was a virtue under the old dispensation, but, when it is done at the expense of the community, may become a crime. Mistakes can be made on both sides; there are misunderstandings and conflicts; but, as Lenin said: 'Mistakes are inevitable – no human being is perfect. The important thing is not to make no mistakes, but to make only few and small mistakes and to learn from them.'

The acquisition of social responsibility has already taken one generation of education and experience, and the old habits still die hard. Nevertheless, it is a step that, in a modern industrial scientific world, must be taken, not only in the Soviet Union but everywhere. There is no chance of turning back. Organization is a necessity to any modern industrial State, but without responsibility this effectively puts the fate of humanity into the hands of plutocrats and gangsters, of whose rule we should by now have had enough experience.

358. One problem facing the Soviet authorities has been illiteracy. When the people of the Carpathian Mountains came under Soviet rule, a total of 70 per cent were illiterate. Courses such as that shown here have been held regularly.

EDUCATION

The social and educational tasks the young Union faced were as great as the material ones. A backward, class-divided, superstition-ridden mass of many races and languages had to find the way to bring itself up to the highest technical and cultural level. The deeply-rooted interests and ways of thought of earlier, capitalist and feudal times, the prejudices and antagonisms, the suspicions and fears that kept man from man, had first to be held in check, then removed by education. In its quantitative aspect the history of education in the Soviet Union has been one of rapid and steady increase, starting even before the troubles of the Civil War were over and still continuing. In 1913 only 20 per cent of the children received elementary education; by 1932 it was universal and the prevailing illiteracy had been wiped out. In 1913, also, only 1 per cent received higher education, and it was confined to the boys of the upper classes; in 1963, 14 per cent received some form of higher education, drawn from every occupational group and with almost as many girls as boys.

The educational work of the Soviet Union was not limited to children, but was also concerned with the grown men, and even more the women, who had so much to contribute to the new civilization. The two great human achievements of the Soviet Union in its early years were the liberation of women from a state of subjection and the making of knowledge and opportunity available to all, irrespective of age, class, or nationality. This is something far greater than the public education of capitalist countries, where the majority are educated for service and only the assimilable *élite* are given the higher education designed to fit them to join the ruling classes. In the Soviet Union Lenin's phrase runs: 'Every cook must learn to rule the State.'

The character of Soviet education has undergone several changes in its short history. For the first ten years after the revolution, the desire to break away from the fetters of the old régime, combined with the feeling that democracy should extend to the schoolroom, led to every kind of experiment in freed education. A few of these had a lasting influence, notably that of Makarenko, who made out of a colony of delinquent children, by emphasizing their self-respect and mutual

359. School education has also expanded in the Soviet Union, with a proper stress laid on science. A botany class in the laboratory.

loyalty, a triumphant and self-supporting commune, whose achievements have been immortalized in his book and the film *The Road to Life*.[7,102] Though Makarenko himself had a tough fight with educationists who could not understand the need for discipline, his work has inspired a whole new generation of Soviet educationists. His fundamental principle, 'the utmost possible demands on a person, but at the same time the utmost possible respect for him', is the keynote of the new attitude towards personality and responsibility in the new socialist world. The inner character of his collective, the Gorki colony, is explained in his own words:

> With this collective there is a very complex system of subordination, each individual personality must harmonize his individual strivings with the strivings of others; . . . in such a way that his personal aims do not become antagonistic to the common aims. . . . This harmony is characteristic of Soviet society, for me the common aims are not only the principal, dominating ones, but they are also linked with my personal aims.

This is an expression of what has been learned by experience in many other Soviet enterprises.

There were, however, relatively few teachers who had the intuition and the character to run schools on these lines. Attempts to do so had in practice a disastrous effect on scholastic standards. In 1928 the pendulum, under Stalin's conservative influence, swung the other way, and discipline and respect for teachers was enforced, together with a heavy curriculum differing from the pre-revolutionary one in a far greater emphasis on natural science. Social education and many play activities were catered for outside the schools by organizations of pioneers and children's palaces. Now a new radical change is under way, directed against the formality and academic character of education and the amount of book work required, and in favour of shorter hours and more practical experience. Similar reforms are going on in universities and higher technical schools.

AN EDUCATED PEOPLE

The educational policy of the Soviet Union is of decisive importance in its whole economic as well as cultural evolution. Its crowning achievement has been the move towards universal higher education. Already, science and culture, from being the possession of an *élite*, is coming to belong to the whole people. It is also becoming evident that a modern industrial State can function and be appreciated only by a completely educated people. The requirements of automatized industry, scientific

360. A recent notable experiment in Russia is the 'Science City' of the Siberian branch of the Academy of Sciences of the USSR. It is built on the Siberian plains in the Novosibirsk region.

agriculture, and research demand a proportion rising to a tenth or more of men and women having scientific qualifications. All must know enough to appreciate and to use science. In turn, only such a wealthy society could afford to keep so many of its population at school for so long.

Nor will the effect of this new universal scientific education be limited to the Soviet Union.[7.87] It is an advance of a kind that will inevitably spread, and not to socialist countries alone; other countries, despite their different social systems, will try to emulate it. In the long run, and even in the short run of a couple of decades, not only dominance but even survival in the world will demand a large output of technically trained personnel and a whole population knowing enough about science to co-operate actively with them. Already, both in the United States and Britain, the recognition that the Soviet Union is turning out from twice to four times the number of scientists per head of population is causing alarm and belated efforts to emulate it. In Britain this has come up against a stratified educational system which has always been considered an essential bulwark of upper-class domination (pp. 1149 ff.). It remains to be seen whether it will prove so strong as to risk the loss of the country's industrial position. Whatever the fate of individual countries, the general tendency is now plain. The world is going scientific, and the sooner people see this and act on it the better.[7.132; 7.146]

SOCIAL EXPERIMENTATION

For reasons already given (pp. 1177 ff.), it is evident that the greatest contributions of the Soviet Union to social science have been in the various aspects of the building of a new civilization, especially in economic planning and in education, rather than in the elaboration of formal social theory. That can only come later when the advances already made can be appreciated in an atmosphere less exposed to stress than that of the Cold War. Meanwhile, it is worthwhile taking into account another aspect of Soviet life, namely, social experimentation and the place of individual initiative. A common mistake, which any experience of the Soviet Union would dispel, is that what has been done there has been done through the directive influence of an all-powerful State and that the role of the people has only been that of obeying orders more or less unwillingly. The reality is very different. In the face of the difficulties, external and internal, already discussed, the achievements of the Soviet Union have, in the last resort, depended on thousands of individual and group initiatives on different levels more or less in the same direction. The role of the party and the government has been to select and foster those it judged, within the limits of its wisdom, to tend towards the general welfare, and to publicize them so that they could be followed everywhere. The whole history of the Union is one of the evolution of new forms of social organization, which began with the *soviets* themselves – spontaneously locally elected councils of workers, peasants, or soldiers with executive powers. More recent have been those of the worker rationalizers, who combine to propose methods of improving work in detail, and worker innovators, who suggest radical changes. In the absence of the old division of interests between masters and men comes the knowledge that a contribution from one is in the service of all. There is no longer any advantage in monopolizing knowledge or skill. We are beginning to see the results of freeing men and women to work, not for a bare livelihood, not for individual gain, but for the community.

FROM SOCIALISM TO COMMUNISM

It is impossible to understand the new civilization of the Soviet Union merely in terms of its history or of its present state. For it is essentially a society with a purpose, that of achieving within a lifetime the complete transformation of which the revolution of 1917 was only the first step – the transition from *Socialism* to *Communism*. The Soviet peoples, undeterred by the threat of the Cold War, are again taking up this task which Hitler's aggression had forced them to postpone. The economy of communism is not something that can be established by a simple act of will. Between the stage of capitalism, 'to each according to his wealth', and the communist 'to each according to his need', there must necessarily be a socialist stage: 'to each according to his work'. It is only in that way that, starting from the inequitable and distorted productive system of capitalism, the productive machinery capable of providing all those needs can be built up. Marx had long ago foreseen the necessity of this intermediate socialist stage. He realized what the utopians of his day did not: the need of a society to be materially rich before it can afford the more generous scale of distribution of goods and services that full communism demands. In his *Critique of the Gotha Programme* (1875), Marx indicated with great clarity the character of communism and the conditions that would have to be met before it could be achieved:

In a higher phase of communist society, after the enslaving subordination of individuals under division of labour, and therewith also the antithesis between mental and physical labour, has vanished, after labour has become not merely a means to live but has become itself the primary necessity of life, after the productive forces have also increased with the all-round development of the individual, and all the springs of co-operative wealth flow more abundantly – only then can the narrow horizon of bourgeois law be fully left behind and society inscribe on its banners: *from each according to his ability, to each according to his needs.*[7,109]

How long that transition to communism will take will depend on the degree to which the peoples of the Soviet Union are left to work out their own constructive purposes without destructive interference from the capitalist part of the world.

With the added experience of nearly a century it is now possible to see what steps need to be taken to make such a transition possible. They were discussed in Stalin's *Economic Problems of Socialism in the USSR* and show how his vision remained clear however distorted his practice. In essence they are the production of *abundant material wealth* and the *rapid cultural advancement* of a socialist society. Only a really wealthy community can afford to be communist, can afford to give away its wealth where it is needed and not to use it to induce some people to produce more than others. Only a cultural community can know how to produce wealth in abundance and to use it.

These are not mere aspirations, but part of a programme that is being carried out stage by stage. The educational developments already referred to show that the second condition is not left till later, but is being met at the same time as the transformation of industry and Nature is being undertaken. The future communist state is felt to be something worth working for here and now. It is no longer a distant Utopia, but a visibly obtainable arrangement of society, the way to which can be charted with ever-increasing precision. It gives to all the peoples of the Soviet Union something to work for that satisfies not only their material needs, but also their sense of justice and human dignity. And they can work for it in the confidence that they or their children will achieve it in reality. Nor is its importance limited to the Soviet Union. What they have done already has stirred the people, the oppressed and working people, of the whole world. The Soviet peoples have shown a way forward and proved that it can be followed. The more closely they approach their goal, the higher their effective standard of material and cultural life, the more certain it is that they will not be alone.

THE PEOPLES' DEMOCRACIES OF EUROPE

For many years now the Soviet Union has not been the only socialist country in the world; it has been joined by others, first in Europe, then in Asia and now in America as well. Between the wars the countries of Eastern Europe, for the most part new states carved out of the old Turkish, Austrian, German, and Russian empires, were in the hands of a land-owning and commercial oligarchy deeply committed to foreign capitalists, first Anglo-French, then German. They were absorbed or conquered early in the War or pressed into Nazi service for the invasion of Russia. When the German armies were rolled back by the Red armies aided by the partisans those countries were liberated from both the Nazis and the old ruling class, who had collaborated with them.

The development of the countries in which the people's democratic governments were formed has, except for an early period of mixed government, been mainly directed by Marxists and consequently has followed fairly closely that of the Soviet Union. All the problems that faced the early Soviet State were here also: devastation, hunger, a small and divided technical and administrative intelligentsia, the still-living tradition of capitalism in the traders and big farmers, together with the influence of the priesthood identified with the old ways of life.

Nor were these the only problems. The policies of the Cold War, which effectively cut off these countries from their former channels of trade, dislocated their pattern of production and deprived them of raw materials which the Soviet Union, occupied with her own reconstruction, was for long in no position to provide. Later the pressures of the Cold War demanded the achieving of economic independence in the shortest possible time. As many of the countries in Eastern Europe had in the old days been essentially semi-colonial producers of raw materials with an industry limited to mines and consumption-goods factories, this meant effectively building up heavy industry from scratch, using for the most part local materials and re-trained peasant labour. The rate of investment was high, and this inevitably reacted on the rate of improvement of the standard of living. There were great difficulties due to lack of experience and equipment as well as rearmament and increase in military forces. Added to this was an external campaign of sabotage and propaganda manned by *émigrés* and lavishly financed by the United States, which magnified every mistake, stirred up every movement of opposition. And mistakes were made: some inevitable, some due to unwise interference from the Soviet Union, itself recovering from the War (pp. 1174 f.). There was an insufficient realization of national and cultural differences, and exaggerated suspicion of political loyalty, which resulted in persecutions and judicial crimes.

When all this has been said, however, the actual progress made in the Peoples' democracies in the few years of their socialist régimes, is most impressive. Cities were rebuilt, factories re-equipped, new industrial centres were created, agriculture largely mechanized, health services increased and the education system extended and thrown open to all. Scientists have for the first time enjoyed the chance of space, equipment, and students, limited only by the speed at which they could train new workers. The breakdown of class barriers, particularly in education, has liberated a wealth of new talent and capacity which, given time, should repair deficiencies of administrative personnel.

CULTURES IN EASTERN EUROPEAN COUNTRIES

It is now evident, especially in the last few years, that characteristic patterns of culture have been developing in the countries of Eastern Europe. The principles of the basic economy are Marxist. It is still, however, a socialist and not communist economy which has been grafted and, in spite of earlier set-backs, it now appears has been successfully grafted on to the original capitalist economies of those countries. In fact, the assertion of national cultures has been rather the feature of the last few years. This applies particularly to the social sciences. There has been much questing on the basis of economics, particularly the economics of agriculture. In Poland, especially, there is a vigorous school of new Marxist economists who are attempting to adapt some of the methods of recent capitalist economic techniques to socialist conditions. In Hungary and in Rumania also there is a vigorous and critical cultural life. The attempt to form an economic co-operation organization in the socialist countries (COMECON), parallel to the Common Market of the capitalist countries, is bound to result in further discussions and in a mutual influence on the economic thought of Eastern Europe and that of the Soviet Union, and, further, on that of the dissident member, Yugoslavia. The large part that Czechoslovakia and the German Democratic Republic, for instance, have played in the economic build-up of the newly-liberated countries of Asia and Africa has brought much understanding of the principles of their development and of world trade.

Most of the countries of Eastern Europe have now achieved an effective economic independence, that is, they can either produce for themselves or exchange with others to secure the continuation and expansion of their national economies. They are, to vary an expression, positively interdependent, and their trade with the western world is also

361. In Hungary and Rumania there is a vigorous cultural life. One aspect of this is the Hungarian cinema, noted for its originality in handling its themes, and for the excellence of its photography. An example is Miklós Jancsó's *The Round-Up*, in which the director brilliantly treats a historical subject – the elimination of the brigand bands of impoverished peasants in the period following the unsuccessful rising against the Habsburgs in 1848. The film is no simple historical narrative but, by a stylization in the handling of its material, provides a comment on the psychological evils of oppression. Brigand betrays brigand and, in the disturbingly tense atmosphere, the successful oppressors are themselves irreparably corrupted.

steadily increasing. That all this has been achieved in Eastern Europe is a demonstration that, in spite of apparently overwhelming difficulties, the enormously increased resources which are released by calling on the whole population instead of only a cultural and technical *élite* can both transform backward economies and raise established industrial states to higher levels of science and production.

THE CHINESE REVOLUTION

For the past fifteen years the transformations in the Soviet Union and its associated countries in Europe have been matched in Asia by the building of a socialist People's Republic of China. The two governments have been welded together in their common struggle against imperialism. This co-operation in the building of a new socialist society exceeds by far the differences that have recently been demonstrated between their two countries on the mode and on the priorities of their tasks.

The new régime in China was one long prepared, with its roots deep in the tradition of the Chinese people. It is a logical development of the three principles of the liberator, Sun Yat-sen. For over thirty years, in struggles first against imperialism and the Chinese feudal war lords, then against Chiang Kai-shek and his betrayal of the Revolution, then, after the Long March, against the Japanese, and finally in 1949 in the victory over Chiang and his American backers, the Chinese Communist Party and its allies have learned the task of self-government the hard way. They learned in these years the needs of the people and the extent of their courage and tenacity.

Now at last, despite American intervention, the people have been given the chance to work out their own salvation. And they are doing it with a will and a cheerfulness that no trials have repressed. Here is not the place to enumerate the achievements of the new Chinese government, but it can be said that no country has achieved such vast and beneficial changes in such a short time. Already China has a rapidly expanding heavy industrial base depending on abundant iron ore, coal and oil. This base is potentially autonomous, but it made at the outset full use of the technical and material assistance that the Soviet Union so generously supplied. The rapid building of roads and railways is ending the curse of poor communications. Great waterworks and irrigation schemes will put an end to the age-old twin dangers of drought and flood, though three years of unprecedented national disasters (1959–61) very much hindered the work of reconstruction.

362. In China industrial expansion is also developing. This 12,000 ton hydraulic press was built by Kiangnan shipyard, Shanghai.

In its social aspect, the Chinese experiment is of enormous importance because here, under conditions very different from those in Russia, the peoples of the most populous state in the world have to discover and create a new and viable social form to fit it for the attainment of independence and prosperity. The cultural achievements of the new Republic have been astonishing, considering the usual slowness of cultural growth. It can be explained only in part by the fact that here the revolution did not so much create as liberate the fund of learning and art, latent in the age-old civilization of China, but held down for years by foreign domination and native corruption. The majority of the intellectuals of China – and certainly all the scientists I have met there, who have preserved through all the bad times the heritage of Chinese culture and are blending it with modern sciences – are enthusiastically supporting the new régime. They are finding a new satisfaction in working for the people and bringing to them the culture that had for thousands of years been the monopoly of aristocratic families.

The transition, especially in recent years with its emphasis on collective rather than individual achievements has not been without its difficulties. Indeed, it can hardly be expected that the ideal of a restricted highly educated *élite*, closely bound up with the Confucian traditions of the Mandarinate, can remain as that of a population where all are educated. Nevertheless in the transformation the best of the old values are being preserved.

Many of the leaders of the revolution were themselves distinguished scholars. Mao Tse-tung is a poet and philosopher. His studies of Marxism and the use to which he put them show how well he understood the way to adapt the teachings of Marx and Lenin to the conditions of a semi-colonial country. Like Lenin he appreciated and experienced the immense value and strength of the common people.

One of his closest colleagues, the historian and playwright Kuo Mo-jo, is the President of the Academia Sinica, and is in himself a guarantee that the best of the old culture will continue in the new. China's past indeed is now more living than ever before. Archaeology has been given a new impetus, and many unsuspected relics of the past have been brought to light. With this goes a widespread effort to bring home to all the people, and not merely to scholars, the value of their cultural heritage.

LITERACY, EDUCATION, AND HEALTH

Already the practical expression of the new cultural movement is making itself evident. The first requirement, literacy, is being tackled with such energy that we may expect within a few years that all but a few per cent of China's population of over 600 million will have learned to read and write, and that within a generation the majority will have secondary education. In the process so far there has been a great simplification of the beautiful but complicated script. At a later stage it is already planned to replace it by a Roman alphabet, thus ending the burden of reading and writing which had been imposed on generations of scholars.

Without even waiting for literary education, dirty and smelly Chinese cities and villages have been drained, cleaned, and brightened as the result of popular health education (p. 980). Inoculation and other health measures will soon wipe out endemic plague, smallpox, malaria, and filariasis. This in itself will release a new fund of human energy for reconstruction. What is certain is that, with the advent of a capable, progressive and fully independent government in China, the world, and particularly Asia, has received an enormous impetus to constructive policies and away from the old cycle of colonial imperialism and war. How that cycle can definitely be broken depends on factors which go much further than the People's Republic of China itself; it depends effectively on whether the processes of national liberation can be completed, bringing the end of colonialism, and whether in that process nuclear war can be avoided and disarmament achieved.

It is on difference of interpretation as to which of these processes comes before the other that the great dispute between the Soviet Union and the People's Republic of China has centred. We must hope that it will be settled by reasonable discussion and not by further bickering and the splitting of the socialist world.

13.7 Towards a World of Freedom

The great political event of the second half of the twentieth century, as the Soviet Revolution was of the first half, has been the rapid liberation of nearly all the colonial areas of the world, that is, the areas that were directly or indirectly part of the old colonial empires of Britain, Spain, France, and Holland. This liberation has so far been brought about in a large variety of ways, from the relatively painless way for Britain in which India achieved her independence to the long-drawn-out wars which were needed before France would relinquish her hold on Indo-China and Algeria. In every case, however, the movement of liberation arose from the people themselves, led by men and women who had long been conscious of colonial oppression and who had seen something in the outside world, especially during the critical years of the Second World War, of the underlying weakness and corruption of the forces of colonial occupation and had learned how they had been shaken and could be again. In this they had the example of the Soviet

363. The emerging countries of Africa are well aware of the need for education and especially higher education. This photograph of the construction of the new Lagos University was taken in 1964.

Union and China before them, not only in the acts of liberation but in the possibility of people building their own civilization by their own efforts.

But liberation has also been a very partial process: independence has so far been largely political and not economic, and by various devices the old and new imperial powers have been able to maintain or reimpose an effective dominance through their control of markets and through investments. Nevertheless, there is no mistaking the general feeling of liberation that has spread all over the world and with it the corresponding feeling of national consciousness and the assertion, now at last in practice, of human equality.

Even in the greatest of industrial nations, the United States, the liberation of Asia and Africa has roused the descendants of the slaves, legally liberated a hundred years ago, to assert their rights more vigorously and effectively than ever before. It is one thing, however, to assert these rights and another to achieve them in practice. The full liberation of formerly colonial countries and peoples, which represent between them two-thirds of the world's population, is likely to be the hardest task that humanity has ever undertaken. The territories and peoples involved are necessarily extremely varied in their stage of evolution. They include ancient civilizations that have recently escaped from the colonial yoke, such as India, Indonesia, Egypt and the Arab States, and those of the populations emerging from earlier forms of clan society in Asia and Africa, as well as those of the Spanish and Portuguese Empires in the Americas, some largely settled from Europe, as in the Argentine, others with considerable indigenous populations, as in Mexico and Peru. Theoretically, the Latin American countries have independence, but economically they are just as much dominated by mining and plantation interests as most parts of Africa today.

Though none of these countries, except Cuba, is communist and in some communism is still proscribed, their people and governments are coming to understand that real independence and prosperity depend on the development of industry and the mechanization of agriculture, and must be built on a basis of education and science.

Already in the nineteenth century Japan had been the pioneer of such developments in Asia, but undertaking it in a period of the dominance of capitalist models, the result was a class-ridden, militaristic state that brought about its own Nemesis in the Second World War. Under American occupation capitalist forms remained, and with them exploitation and poverty. On the other hand, the Japanese people, long suppressed, are beginning to assert themselves and to demand that they should shake free from US domination and be no longer isolated from China, but should, for their general benefit, co-operate in the renovation of Asia.

In India, liberation from British rule has been followed by developments which, though they do not follow the same pattern as in China, are inevitably moving in the same direction. Though parliamentary democracy is maintained in the political field, economic policy is turning towards planning and limited socialism. A limited amount of success has been achieved in making and partly carrying out three Five Year Plans for economic development, but these plans have been hampered at every turn by the interference of foreign interests, larger now than before liberation, working in alliance with well-developed native capitalism in the so-called private sector of the economy. These difficulties have been further aggravated by the tragic Chinese–Indian frontier dispute, which has brought India partly into the orbit of the Cold War and added the military to all the previous burdens on its economy. Whether it will be possible to arrive at a solution of India's economic problems and provide for the people's welfare in the orbit of a capitalist economy is most disputable; in so far as it is not, it will be for the Indian people to get for themselves an economy that will do so.

However that may be in the future, India has already had very considerable effect on international politics, wielding its influence as a large neutralist state which favours solutions on international disputes by means of negotiation under the auspices of the United Nations. It has firmly established the existence of a neutralist group of nations which has done much to mitigate the hostilities of the Cold War.

THE ARAB STATES

The group of states stretching from Pakistan through Iran to Morocco formed the central part of the old Islamic Empire and then fell successively under Turkish and latterly British and French domination. During this century, more or less violent struggles for liberation have established them all as independent states, but it is still too soon to measure how far they can achieve effective economic independence. With neglected economies and superficially poor natural resources, they have had a hard struggle. Their real natural resources are, however, enormous, particularly in oil and ores. Their problems are essentially problems of liberating the ordinary people of those countries so that they can take possession of their own resources and use them for their own benefit. The problem is also largely educational: popular education requires the effective supersession of a small landlord and trader class, which will not be achieved without a struggle.

CENTRAL AND SOUTH AFRICA

The liberation of central and south Africa came so rapidly that the people themselves were often surprised by it. Africa, which was carved up between the great powers in a decade at the end of the nineteenth century, became free in an equally short time in the twentieth; but that freedom, as the case of the Congo in particular has shown, is a very qualified one. Indeed, the policies of the imperial powers of withholding higher education from the inhabitants of colonial countries meant that there could not be an efficient force of administrators and technicians at the moment of liberation. This was indeed what even the most liberal colonial governments wished. They intended to keep the peoples of those countries in permanent dependence, but the tempo of liberation was too fast for them.*

The social problems arising from African liberation are as great as the material problems already treated (p. 964). Essentially they

364. Oil resources in the Arab states are enormous. An oil pipeline in Iran.

consist of canalizing the spirit of liberation in the direction of training the people to take charge of their own countries for their own benefit and to get rid of the tutelage as well as the exploitation of colonial powers. Under the old régime their whole economy had been distorted, with the object of providing raw materials from mines and plantations, and the development of local industry had been restricted to processing. Administration, also, except for that by the lowest functionaries, had been in the hands of foreign powers and still is in the many French ex-colonies. To remedy this in the long run is a matter of general education, but in the short run it is necessary to work with whatever means are available, including a certain number of European teachers and administrators. The difficulty here is to prevent those persons operating, consciously or unconsciously, in the interests of their home countries and at the same time to retain their effectiveness.

The whole of Africa, as well as other similar areas of ex-colonial territories, is exposed to the pushes and pulls of the Cold War in the most intimate way. The conflict between private and public interest is one of these areas. Any move towards socialism is deemed to be in the interests of the Soviet Union, and, similarly, predominant concern with private enrichment, in those of the United States.

It is in just such a disturbed state that there is a possibility of building up a new social consciousness, expressed so far in very vague and even mystical terms such as Africanism or the African personality. But in one way or another this problem is going to be solved and the result will be something infinitely different from what emerged in Europe or even in parts of Asia. To get education working, for instance, on a sufficient scale to provide the necessary administrative ability and practical engineering capacity may require much simplification of European traditional methods of education, most of them largely concerned with keeping the number of specialized candidates down in order to secure adequate remunerative employment for them. In a world at peace, or at any rate one of peaceful co-existence, this problem will be much easier and might be solved by rational international co-operation under a United Nations that was really representative, which the present one is not. This is a problem that is already occupying many of the scientists, social as well as physical scientists, in the older countries. Nevertheless, they are mostly not yet sufficiently aware of the unconscious relics of the colonial attitude in its most beneficent mood, where the help carried with it an implication of cultural superiority which, for instance, was already implicit for years in the missionary movements which paved the way for the conquests of colonialism.

LATIN AMERICA

The countries of central and south America formed part of a frankly exploitative Spanish and Portuguese empire, composed at the top of slave-owning planters employing Negro or native Indian slaves on big and badly managed estates and mines which were the property of the crowns of Spain and Portugal. The first liberations at the beginning of the nineteenth century got rid of this latter situation by handing over royalties to big landowners but changed little else. Slave-owning in Brazil was abolished only in 1888 and raw naked exploitation was largely replaced by a much more efficient and humiliating exploitation by the industrial monopolies of Europe and the United States, which used the Americas, as they did Africa, essentially as sources of raw materials.

There was, however, an enormous difference from Africa, through the presence of a large and educated emigrant population, especially in the states of the southern part of the continent, Argentina, Chile and part of Brazil. This generated a vigorous Latin American intelligentsia which found its outlet largely in politics and in the arts, a transplantation of eighteenth-century Spanish civilization to the New World. Because industry, other than mining, was little developed and because most public enterprises were in foreign hands, the material basis for natural science was largely absent. But this situation is changing very rapidly with the greater consciousness of the economic dependence of Latin American countries on the United States, in particular, and on other capitalist countries. The one example of Cuba, which has broken completely from the old colonial and neo-colonial system, has had an enormous effect on the intelligentsia as well as on the peoples of Latin America. The continent is a very rich one, rich in human as well as material resources, and, although it is too soon to see it yet, they will certainly have a great contribution to make to the development of social sciences in a newly united world.

13.8 The Future of the Social Sciences

The contrasted picture of two social sciences, one drawing its inspiration from Marx. the other linked with the fate of capitalism, is characteristic of our divided world. But it does not, as I have already explained, imply the perpetuation of this division, nor mark an abyss that the social thinkers on one side or the other cannot cross. In the first place Marxist thought is by no means limited to the new socialist countries, but is being developed in every country in the world today. Further, workers in the social sciences, just as those in the natural sciences in all countries, are heirs to a common tradition which they may interpret in different ways, but which should enable them to understand each other.

When we talk of the future of the social sciences, that future must be seen against the social background of the future of civilization. If another annihilating world war can be avoided – and the peoples of the world should be able by their united action to prevent its outbreak – we may expect for many years the continued existence of two contrasting economic systems. To maintain that co-existence will involve a great increase in trade and cultural exchange between the systems. This should remove at least the grosser misconceptions of social thinkers about their colleagues of a different persuasion. It cannot and should not lead to uncritical acceptance on either side of the ideas of the other, but to a criticism carried out rationally and not with atomic bombs.

The criticisms made here of the social sciences in capitalist countries are not aimed against those who are honestly trying to probe deeper into human relations in society. They strike rather at the system which, almost automatically, distorts and perverts all such pursuit of knowledge to serve limited and mean ends. Just because the social sciences deal more directly with human life than do the natural sciences, they are, for the moment, in the countries of Western Civilization, more securely tied to the defence of privilege and the preparation for war. Nevertheless, sooner or later the growth of the social sciences will have effects that their promoters have not intended. It is impossible to use, still less to develop, any science without bringing out the latent possibilities for fundamental criticisms which it contains. It will be in the struggle to liberate their disciplines and to bring out the implications of their criticisms that the new social sciences will find their appropriate forms.

The world outlook of the capitalism of today is not one of aspiration to a brighter future but one of desperately clinging to present inequalities that can be glossed over but cannot be indefinitely preserved. In so far as the social sciences reflect the values of capitalism they are inevitably bound to regress. They may well go even farther in their apologetic and mystifying role, and add many new chapters of statistics and logical and psychological analyses, but of no fundamental im-

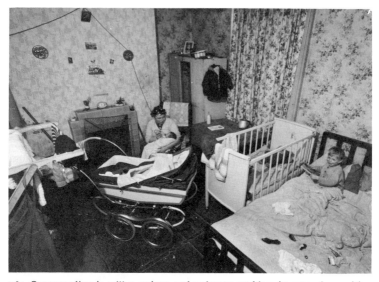

365. Overcrowding is still a serious and unhappy problem in many large cities. Here a married couple and their two children are living in one room, sharing a kitchen with two other families. Their only running water comes from a bathroom which they share with others in the house. The ceiling of their single room is in bad repair and likely to fall in.

portance. But they need not, even in the capitalist world, reflect these values. With the release from the stranglehold on ideas of the Cold War there will be everywhere a greater freedom of thought; indeed it is already beginning to be felt. We are entering a period of critical re-examination of the basis of society and of human behaviour in the socialist as much as in the capitalist parts of the world, and we await the contributions that will come from Asia and Africa. Undoubtedly, the influence of Marxist thought will make itself felt, but it will be increasingly so on account of its intrinsic merit, and it need no longer be accepted or rejected unthinkingly. Nor need we fear that any of the valuable traditions of past learning will be lost. Here in the ending of the great schism of human thought will be work enough for generations of humanists and social scientists. For my part I can only understand the world as I have learned and experienced it, that is, largely in the light of Marxism, and my picture of its further development must lie along the same lines. But I know also that this is only a short-period

view and that in time deeper and more comprehensive interpretations
will come.

A NEW SCIENCE OF HISTORY

It is only when Marxist criticism is brought to bear on history that the
multiple and confused stream of events in human societies begins to
acquire a significance far greater than was apparent to the classical or
the scientific historian, or even to the people of the times described.
History now shows an intelligible pattern, and is not just 'sound and
fury signifying nothing'. Great events like the Renaissance, the Reformation, and the French Revolution become more fully intelligible and
fall into place as phases in a more extended class struggle.

Even the history of the ancient world, which has been the dullest and
most severely limited field for pure scholarship, appears as a lively
scene of social conflict, bringing out in a simplified form many of the
economic factors that are difficult to find in the greater complexity of
our modern civilization. It is possible to show in classical history that
the whole of civilization, not only the political and economic events
but also the philosophy and the aesthetic achievements, is all linked
together in one unity, and thus all acquire a deeper and more immediate significance for us today.[7·52−5; 7·162; 7·163]

In the light of materialist dialectic the motives and movements of the
past come to life. The very controversies that its introduction has
stimulated have contributed to new researches into facts and a sharpen-
ing and clarifying of arguments. Indeed already the work of Marxist
historians is felt all over the field of history. The only way to escape its
influence is to reduce history to personal trivialities or to return to a
divine providential view[7·21] that was outmoded even in the eighteenth
century and that is not taken very seriously today, even by professed
believers.

An enhanced significance is also being given to the auxiliary sciences
of archaeology, philology, and anthropology. The nineteenth century
was a great period for discovering the remains of the past, the habits
of existing primitive people, and the enormous range in space and time
of human languages; but as long as simple description was the only
discipline that could be used these remained interesting but meaningless.
The clues to the study of man and his evolution are the clues of social
and economic factors, of conflicts and of the splitting of homogeneous
societies into classes giving rise to new social and economic forms. The
whole range of human history, stretching back to the very origins of
humanity, now begins to acquire an importance of a different order

from that which it had in the past. It is needed in understanding our present society and to mould a future one.

A NEW PRACTICAL SOCIOLOGY

Sociology, which is the anthropology of modern societies, is the science most closely tied to the capitalist machine. Its function has been to analyse social, political, and industrial situations in order to secure the easier running of the system. It has also had the task of explaining the system in a way that will justify and even do it credit. It had therefore above all to avoid serious criticism. Yet it is precisely in the direction of such criticism that lies the only hope for sociology, if it is to be considered a science and not mere window-dressing. Indeed it is not difficult to find subjects for criticism; what is difficult is to find the courage to make the criticism and any means of getting it published. But the honest sociologist has an ally much stronger than his present employers. The people know well enough that they are being held down and cheated. What has not been so clear is how the system operates or what can be put in its place. In the future, as in the past, the people will support those who can help find these things out and explain them simply and clearly.

PSYCHOLOGY IN A CLASS SOCIETY

Of the future of psychology it is not necessary here to say much. Certain parts of the present area of the subject, such as the psychology of sensation and nervous control, belong essentially to biology and may well advance along existing lines, though here far more attention will have to be paid to the work of Pavlov and his school. The major part of psychology is only the reflection of society on the lives and feelings of the individuals who are shaped by it and in turn contribute to changing it. Present-day psychiatry in capitalist countries may, by analysis, drugs, or operation, palliate some of the ill-effects of society on the individual; but this cannot touch the causes of the trouble. The science on which these practices are supposed to be based shows all the signs of suffering from the same limitations. It cannot escape them without rejecting the deep-seated errors which arise on the one hand from abstract reasoning based on individual introspection, and on the other from purely biological analogies.

APPLIED SOCIAL SCIENCE: LAW

The whole group of practical social sciences – law, politics, and, youngest of all, economics – were themselves social products, and have in their times shown a vitality corresponding to the importance of the social function they fulfilled. But under capitalism, in all cases, that function has tended to become a negative one. Law is, as we have seen, completely parasitic on a political and economic system, so much so that it is not surprising to find the most fanatical defenders of the system, either in its naïve 'conservative' or its sophisticated 'labour' form, among the lawyers. Law, in fact, is so wrapped up in property and property relations that in its present form it can hardly survive the alteration that a genuine economic revolution would bring. It has accordingly become one of the most arbitrary and unscientific branches of social studies. In law, more than in any other social field, it has been advisable not to look too closely at foundations, not to attempt any rationality or scientific approach, and never to experiment.

Marxist criticism of law demonstrates its historically relative and class character, and the hollowness of claims to its absolute and universal jurisdiction. The range of interest of the law had grown continuously with capitalism to cover almost every aspect of social relations. In the Socialist economy it is necessarily different. Civil law merges into administrative law, and criminal law in principle becomes largely a social service for the adjustment of personal relations and anti-social behaviour. Such a transition, equivalent to that achieved in centuries in the capitalist world, has by no means been completed. Indeed, it was blocked because for a large part of the time the Soviet Union was virtually under martial law and the security police were given almost unlimited power, which they used only too often unjustly. These abuses will not, however, now that constitutional rights are being restored, prevent an evolution of law to fit the new forms of society.

POLITICAL SCIENCE AND ECONOMICS

With politics the situation is somewhat different. Political science as hitherto taught has very little claim to status as a genuine science. Political practice under capitalism has been a mixture of secret financial negotiations and open demagogy, most blatant and ill-concealed in the United States, more polite and difficult to detect in the older and more sophisticated capitalisms of Britain and France. A science invented to deal with politics is necessarily under an overwhelming pressure to justify this state of affairs in learned language. It serves largely to perpetuate the illusion that words such as *democracy* and the *two-party system* correspond with some fixed and unsurpassable ideal.

The discrepancy between an ideal theory of politics and the actual practice of politicians leads to a general scepticism of politics. This scepticism can be effectively played on by those politicians who are not in the gang themselves, but who would like to be, and is a fertile source for Fascist ideas. In a society that is, as a whole, seemingly purposeless and effectively contrived to afford the maximum profit to a few, it becomes indiscreet to inquire too closely as to the economic foundations of political action. It is accordingly useless to expect a critical, still less an experimental, political science to be blessed by the authorities. Only in the ranks of those who are fighting against the system can such science be found. To be effective, political science cannot be separated from political action on the one hand and a general economic and social analysis on the other. In the future political science will be neither an academic description of social dynamics nor a manual for successful politicians, but a part of the education and practice of every citizen.

Of the relatively new science of economics we have already spoken, especially of its effective bankruptcy in a period of monopoly capitalism with its alternations of crises and wars. The mechanisms of control of economic factors are not in the hands of the economists, who, for the

366. The world requires technical and scientific education if proper use is to be made of natural resources. The higher schools of the Soviet Union train between 260,000 and 290,000 specialists annually.

most part, have to be content to be the chroniclers of the decay and collapse of the system which they are serving. In so far as that description and analysis is honestly carried out and does not seek to hide the true state of affairs, or to propose remedies which are bound to prove mere short-term palliatives, the economist will contribute to the founding of a better society.

TOWARDS A NEW SOCIAL ATMOSPHERE

The atmosphere that of recent years has weighed heavily on the social sciences can be fully dispelled by socialism and in part by the struggle to achieve it. The illusion and hypocrisy, the brutality and despair that underlies so much of our present-day thoughts, can be turned into their opposites (pp. 1192 f.).

There is no need for illusion once we admit the central task, which is to increase man's mastery over Nature through learning its laws. In carrying out this task man frees his own powers and transforms himself in transforming Nature. This implies that the world is knowable, and through knowledge controllable.

There is no need for hypocrisy once man admits the plain evidence that the society of the capitalist period is an unjust society, and that, whatever merits it had in its formative time, it has now outlived its mandate and is only a brake on knowledge and progress. It is also necessary, as recent experience has shown, to watch the evolution of socialist societies and prevent abuses which may pervert their purposes and destroy their essentially democratic character.

To brutality we must oppose the doctrine of equality of people, of sexes, races, and nations. And this is no abstract doctrine, but a real opening of opportunity to all by means of an education up to the highest levels available to all boys and girls alike and an expanding economic system making ever-increasing uses of natural resources.

This also is the answer to despair: a future free from war because its economic and social roots will be destroyed, free from material anxiety because of the unlimited possibilities that lie in the application of knowledge. The old picture of humanity struggling ever more viciously for ever-diminishing resources, and the nightmare of neo-Malthusianism, will be exposed as mere excuses for the maintenance in luxury of a privileged few. Man can now gain for the first time, without any need for supernatural support, a full confidence in his power over destiny. All that has happened up to now he can regard as prehistory; the new stage of real history is the conscious control of social and material forces by men themselves.

It is in the light of such a prospect that we can begin to appreciate the full importance of the social sciences. In their various disciplines they will in the future have a far greater part to play than in the past. In the process of the transformation of society they will acquire a new responsibility, and will be linked in a way they have never been before with the productive process and with the natural sciences. The social scientists will no longer form a small and separate band, observing but never experimenting, but will work alongside the people controlling and transforming their own society.

THE STATUS OF THE SOCIAL SCIENCES

This long review of the social sciences had as its purpose to round off the picture of present-day science. Though complex and inevitably incomplete, it should serve its purpose if it brings out, more clearly even than in the natural sciences, the link between social knowledge and the state of societies of the past and present. At the outset the status of the social sciences was discussed, and indeed their right to be considered sciences was put in question. It should by now be clearer why such doubts arise. For the study of society is revealed as having deeper roots in the past and a closer association with the dominant social powers in the present than have the natural sciences. The inability to carry out genuine experiments under capitalism is only one symptom of its fundamental arbitrariness.

In so far as the social sciences remain attached to the old social forms they have no future; but once they break away they will find new inspiration and new fields of activity. In socialist and non-socialist countries alike there is a future for the free and critical study of man and his institutions. This study, like that of Nature, will gain its full status in the measure that it is linked with the work of all the peoples to build for themselves a peaceful, abundant, and active life. For this, however, the first pre-requisite is the establishment of a secure peace without which, in an atomic age, civilization cannot advance and where it is doubtful whether it can exist at all. If we can preserve peace, the great task, and one that falls especially on the social scientists, is a re-examination of the basis of society, socialist as much as capitalist, to find whether in a common and constructive effort the profound causes of the conflicts of our time can be laid bare and eliminated. Man has already survived other periods of disaster and of shaken beliefs, and, from the questions these have raised, made the greatest advances in the understanding and mastery of his society.

CONCLUSION

This section completes the seventh part of the book, which deals with science and its social impact in the current century. The division into chapters covering the physical, biological, and social sciences was made necessary by the scale and range of advance in the last fifty years. This separation had, however, the disadvantage of not bringing out the significant inter-relations between the sections nor showing how the whole of the scientific effort of the time is tending towards one general socially evolved means of understanding, controlling, and transforming, stage by stage, the whole environment of man and, in the process, the society he has made and in which he lives. This is a task which will be attempted in the last section of the book.

PART 8

Conclusions

Chapter 14

Science and History

14.0 Introduction

We have now reached the end of the account of the development of science and technique throughout human history, and the time has come to attempt to assess the meaning of the story that has been told. Many of the significant lessons have already been shown in their context in the different chapters. The main weight and value of the argument arises, indeed, directly out of the specific interactions of science and society as they occur. Nevertheless, it may help the reader as well as the writer if we pull together the threads of the past of science, and try to put in a few words what this may mean in the present and the future.

In this chapter I shall begin (14.1–14.3) by discussing the general character of the interaction of science and other social forces as they have emerged from the more detailed studies of the previous chapters. In doing so I shall attempt (14.3) to answer some of the questions raised in Chapter 1 as to the factors which determined the particular association of science with events in different times and countries, with special reference to the great bursts of scientific activity which occurred in Greek times, in the Renaissance, in the Industrial Revolution and, the greatest of them all, that in which we are living today, and (14.4) to the part that class divisions have had in the growth and character of science. This leads me to a consideration (14.5) of the place of science in the present world and to an attempt to answer other questions which are of immediate concern to scientists, such as militarization and secrecy in science, and its place in government. In the next section (14.6) the conditions favourable to the advancement of science are discussed, including the scale of support and the inner communications of science. This leads to the two vital questions of the place of philosophy in science (14.7) and of the apparent incompatibility of freedom and organization in science (14.8). Finally (14.9) there remains the question of the immediate tasks of science and its perspective in the future.

14.1 Science and Social Forces

The story of the development of the social enterprises of science, technology, economy, and politics through the ages does in itself indicate something of the nature of the connexions between them. This can be seen from the tables on pp. 240-43, 336-7, 134-5, 670-71, 864-5, and 990-91. Table 8 (pp. 1312-13) presents the general lines of the development of scientific theory and practical technique from the origins of human society till today. In the most fundamental sense it is a picture of how man got his living, beginning with food-gathering in the wild, in a way differing but little from that of many animals, and ending, for the present, in the mechanical exploitation of multiple resources, in which more and more of the processes of Nature are being directed to human service. Each of the various stages of advance is marked by the introduction of some new technique based on a deeper understanding of the way Nature works. Thus agriculture passed from scattered temporary gardens to permanently irrigated fields with the invention of the wooden ox-drawn plough (p. 115). The axe and iron plough-share furthered its spread to forest lands (p. 149), the reaper and binder, and later the tractor, threw open to it wide, sparsely inhabited dry grass-lands (pp. 653 f.).

Most technical advances are, as we have shown, responses to social economic demands, made, entirely in early times and sometimes even now, by the craftsmen themselves working over and improving their traditional skills. At first there was no place for science as we know it. This emerges in recognizable form from the common social tradition of the crafts, only with the beginning of city life – of civilization. The first appearance of a distinct science was for economic and control purposes: in the numbering needed for temple accounts from which come, by a continuous tradition, all our mathematics and writing (pp. 117 f.). The tradition of numerical and literate science has been passed on from master to pupil for some 5,000 years, broadening and sending out branches wherever there was a call for large-scale order and organization, dwindling to a trickle in a fragmented society such as that of early Saxon England (p. 234).

The two streams of technical and of scientific tradition have, for most of history, run on apart from each other. The conditions of early civilization led to a division of classes which put the scientists among the clerks on the side of the rulers, while the manual craftsmen were only a grade above the peasants and were often themselves slaves (pp.

130, 165 ff.). That division has persisted to this day, and its final breakdown is only beginning. Meanwhile, for long periods of time it condemned science to sterility and technique to repetitive stagnation, while
radical advances came only rarely.

EPOCHS OF CULTURAL ADVANCE

Those opportunities came, as has been shown, in the few periods in
world history when the class distinctions partly broke down and there
was mutual stimulation of scientist and craftsman. Counting as one
period the first creative outburst of early civilization (pp. 99 f.), before
the division between scientist and craftsman occurred, we can find in
the whole of history only five such major periods. The other four were:
that of the Greeks (pp. 159 f.); of the sixteenth and seventeenth centuries in Europe (pp. 373 f.); of the Industrial Revolution in Britain
(pp. 518 ff.); and of the scientific technical transformation of our own
time. Compared with these the Islamic–Medieval synthesis (pp. 270 ff.)
and the various bursts of scientific technical creativeness in India and
China (pp. 262 ff.), though they played essential parts in the transmission of culture, were in themselves relatively minor developments.

Both the major and minor creative periods in science and technology
appear in history as accessory to great social, economic, and political
movements. The first period corresponds to the setting up of the first
States and empires, those of the Pharaohs of Egypt and the kings and
emperors of Mesopotamia. It marks the merging of clans into a class
society (pp. 107 f.), but by no means a complete transformation, as
much of the clan organization persisted for centuries, as it has in China
almost to our own time. The second, the Classical or Greek period,
was that of the growth, triumph and decay of a money and slave economy, marked by open civil and class warfare (pp. 190 f.). After the fall
of the Roman Empire its place in Europe was taken by a different,
much more decentralized and locally self-supporting, feudal economy,
making at first little demand on science and contributing correspondingly little to its advancement (p. 334). The decay of the other great
empires, Persian, Maurian in India, Islamic, led to a similar situation
in most of Mediterranean and Asian countries, though not in China.

The third period of advance, from the point of view of science perhaps the most important of all, was that of the Renaissance, which
marked the beginning of the supersession of feudal economy by a new
bourgeois economy; while the fourth, the Industrial Revolution, coincided with the definite establishment of manufacturing capitalism as
the dominant world economy. Capitalism is distinguished from earlier

money economics by the use of the profits of enterprises as capital for further investment, leading to a rapidly expanding industrial development involving machinery and the use of power from coal. This was the first stage of a transformation of importance equivalent to that of agriculture, for it concentrated the economy on or near the coalfields, and drew the sustenance needed for the vast populations there from all over the world.

The last period of technical scientific advance is that which has now been in full swing for some fifty years. It is marked in the social field by a struggle of unprecedented violence and range between the capitalists and their supporters on one side and the new socialist forces based on the working classes on the other. Through all this, technical and scientific progress continues ever more rapidly in the service of both war and peace. It is characterized by the penetration of science into all forms of production, and by an ever-increasing degree of organization and intercommunication. It is now evident that enough is known, both of natural science and of techniques making use of it, to solve all the major problems of world economy. We should now be able to provide a tolerable standard of life for the peoples of the whole world,

367 a, b. The ancient and time honoured shaduf which could only irrigate a small area, compared with a modern irrigation system. This irrigation canal in the Soviet Union is 170 miles long and waters some 4,700 square miles of once arid desert.

and by further research it could be improved indefinitely. We are certainly not doing so now, and whether we ever do it is dependent on the solution of the social and economic problem, while present inner contradictions and open conflicts waste resources and threaten the extension of war and famine.

THE FUSION OF THEORY AND PRACTICE

In every one of these periods the great advances recorded seem very largely due to a closer fusion of theory and practice than was the case in the intervals between them. It was, as has been shown, in each case the social fluidity of the time that allowed a coming together of craftsman and scholar, sometimes a complete merging of the two. It was on one side the 'Naturalist's insight into the trades', to use Boyle's phrase (p. 461), and on the other Robert Norman's claim that 'yet there are in this land divers Mechanitians, that in their severall faculties and professions, have the use of these Artes at their fingers endes, and can apply them to their severall purposes' (p. 435). This interpenetration has in the past been limited and temporary. After an interval we find that on one hand the scientific theory injected into technique has become traditional craft lore, as for instance in the art of the spectacle-maker, or that on the other the contact with practical things has been embedded in scientific theory, as were the mechanics and pneumatics of the later Greeks. It is only in our own time that we can see an approach to complete and permanent fusion between the scientist, engineer, and craftsman, which can be completed only by the achievement of a completely classless society.

These are the historic facts with which I have dealt in earlier chapters, where I have also tried to explain how the different scientific, technical, economic, political, and ideological aspects are related in each case. Nothing can take the place of a study of the facts far more detailed than any I have been able to undertake here. Nevertheless, on the basis of what has been shown, certain specific explanations begin to appear that throw light on the actual course of history, and certain general conclusions can be drawn which seem applicable to the present and the future. These will be discussed in the following paragraphs.

368 a, b. The hit and miss methods of the alchemist based half upon magical considerations as depicted by David Teniers (1610-90), contrasted with a modern chemical laboratory where techniques and theory combine to provide a real control of nature and natural resources.

THE GROWTH OF SCIENCE

A deeper understanding of the process of cultural development can be gained by looking at the facts in a different order, by following separately the streams of scientific and technical advance. As already indicated, the scientific stream was at its outset a very narrow one, narrower in reality than it was in appearance at the time. We can, in the light of our present knowledge, distinguish what in earlier knowledge was valid and capable of verification and use, from what was vague, mythical, or downright nonsense. Thus we have by now discarded astrology, alchemy, and cabbalistic number lore, though in their time they were all reputable sciences and added much to the prestige and income of the learned. The same may be said for all medical theory and much medical practice up to the nineteenth century and beyond. Nevertheless, bad theory if followed and tested out can lead to good, though with a sad waste of intellectual effort. Astronomy and chemistry are the offspring of astrology and alchemy (pp. 398 f., 410).*

It is, however, round the hard core of verifiable and usable science

that the rest of science has been organized. This is the mathematical–astronomical tradition that has continued unbroken, piling observation on observation and adding method to method, since the time of the first cities 5,000 years ago. The practical difficulties of reconciling the lunar and solar calendars, which are still with us in fixing the date of Easter, proved a testing ground from which the whole of mathematics, algebra as well as geometry, developed (pp. 121 f.),

The Greeks, from distaste for or weakness in numerical calculation, created geometry to give a visual and mechanical picture of the celestial world, virtually the beginnings of mathematical physics. This astronomical core was the centre not only of Greek science, but of philosophy as well. It was extended, as we have seen (p. 308), quite illegitimately to explanations of human physiology, and also served to justify the social hierarchy. Indeed, apart from the image of the celestial world, Greek science consisted only of descriptions and classifications of the world of Nature, the value of which for science and practice was only to be felt in the nineteenth century (pp. 203 f.). Medicine, the other respectable branch of science, falls into this category. Diseases were admirably described and there was good diagnosis and prognosis, but the best treatment was to let the patient alone (pp. 186 ff.).

Astronomy and medicine, often practised together by the same men, saved the essence of Greek science after the break-up of classical civilization and carried it all over the world. Thus for the first time there grew up a common world science ranging from China to Spain. What was done, mainly by the Indians and Arabs, and the Schoolmen (pp. 332 f.) was not so much to add to the central tradition but to tidy it up. It led to a better welding of geometry, algebra, and trigonometry, with astronomy (pp. 303 ff.), The introduction of a trivial, and not very original device–the Arabic numbers–was to have a decisive effect in speeding up calculation and extending its use. At the same time, in those auxiliaries to the medical profession–chemistry and optics–the Chinese and the Arabs made a first break out of the limited range of Greek science (p. 278). The main importance to science of the long gap of the Dark and Middle Ages lay in the nurturing of new technical devices which made themselves felt only in the Renaissance. These included the great and expanding break-throughs represented by the inventions of the compass, gunpowder and printing derived from China (pp. 311 f.).

Nevertheless, in the Renaissance and until the end of the seventeenth century, the chief interest in science remained in the skies. Intellectually the centre of the drama was the dethroning of Aristotle; the destruction

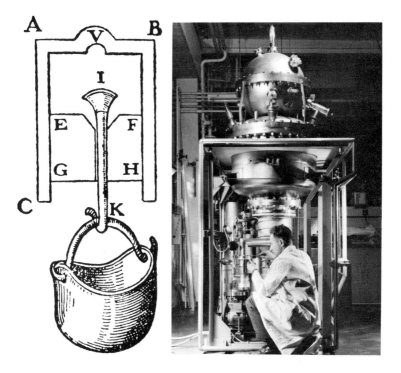

369 a, b. It was Galileo who in his *Discorsi . . . intorno à due nuove scienze*, Leiden, 1638, gave details of a piece of apparatus to demonstrate the existence and power of a vacuum, so abhorred by earlier philosophers. ABC is a section of a glass or metal cylinder with a recess at V. EFGH is a wooden plug, K a metal hook at the end of a thick wire IK. The end of the wire I fits into a recess in EFGH and the plug is pushed up 'two or three finger breadths', the apparatus inverted and filled with water. Upon turning the equipment over so that the mouth of the cylinder is downwards, a vacuum is created when weights are placed in the bucket. It is the vacuum created by the downward motion of the wooden plug that prevents it from falling out. Experiments of this kind were carried out twenty years later by Otto von Guericke in Magdeburg.

Galileo's simple apparatus is contrasted with modern vacuum technology where vacuum pumps producing a very high degree of exhaustion are used, as is the case with this pump and calibration apparatus for producing extra-terrestrial vacuum conditions at the European Space Technology centre.

of the hierarchical, feudal world-picture and its replacement by one based on impersonal natural law. This was the most daring of human theoretical achievements, and though relatively slight in itself, made all subsequent science a straightforward business. The decisive step that was taken in the Renaissance was to give to exact science an immediate practical value, by using the new astronomy for navigation, itself a major factor in achieving world dominance for the new bourgeoisie. The seventeenth century further witnessed the first break of science into the technical field through the discovery of the vacuum, which was to lead to the steam-engine (pp. 469 f., 576 f.).

Only in the eighteenth century did science begin to expand into new fields quite outside those studied by the Greeks. Indeed the main centre of interest shifted from astronomy to mechanics and chemistry, which linked it with the main interests of the Industrial Revolution – power engineering and the textile trade. It was from the latter that was to come the renovation of chemistry with its final break with alchemy and the development of the atomic theory. The great comprehensive laws of the nineteenth century – the conservation of energy and the electromagnetic theory of light – mark the extension of mathematics over the whole field of physics. At the same time the first break is made into the biological field with the work of Darwin and Pasteur (pp. 643 f., 647 f.).

The great advances of the seventeenth, eighteenth, and nineteenth centuries in the practical understanding of the world and its use was reflected also in human thought. The Newtonian concept of the world of law was to have profound philosophical and political consequences (pp. 487 ff.). The materialist and atomic hypotheses restated in a concrete way, on the basis of the Greek imaginings, helped to shake idealistic and spiritual views of the universe; the idea of evolution and its application to mankind itself showed for the first time what was man's place in Nature (pp. 643 f.). Henceforth, no philosophy or religion could escape the influence of scientific concepts (pp. 660 ff.).

In the twentieth century all barriers are down; there is no field of Nature into which science does not penetrate. At the same time science is becoming creative, building up a world of its own of mechanical, chemical, and electronic devices, and their use is tending to replace the fruits of purely technical development.

THE COURSE OF TECHNICAL ADVANCE

In contrast to science, with its narrow competence, technique has always had to advance on a front as broad as the whole pattern of contemporary life. Over most of the time, that advance has been relatively slow, moving forward only when a new material or device opens up regions hitherto inaccessible. Stone, bronze, and iron were recognized from antiquity as marking eras in human culture; they were technical achievements owing nothing to science. The same can be said of the revolutionary introductions of fire, pottery, weaving, the wheel, and the ship (pp. 69 f., 95, 115).

Right up to late Greek times, when most of the techniques on which our lives are based had already been evolved, it does not seem that science, except in such by-products as monumental architecture, and perhaps in water-works (p. 218), entered technology at all. The speed of its advance seems to have depended entirely on social and economic factors. The great jump came with river-valley agriculture, providing a surplus of labour out of which cities could be built and craftsmen fed (pp. 100 f.). We may suspect, though we cannot prove, that the professional scientists had a hand in the development of machinery, in the design of gears, screws, and pumps, even if it was only to polish up crude devices invented and used by the workmen themselves.

Certainly nearly all the set of inventions, mostly Chinese, that were to change medieval into modern economy – the horse-collar, stern-post rudder, trip-hammer, and mechanically driven bellows – owe nothing to science. Even spectacles, gunpowder, and printing are largely practical achievements, though the inspiration must have come from the learned (pp. 320 f.). Only in the compass and the clock, essential for navigation, does the scientist seem to have made a larger contribution (pp. 315 ff.).

What is called the first Industrial Revolution – that of the sixteenth century – is almost entirely the fruit of the skill of workmen under the impetus of the new capitalist system, with its rewards for productive enterprise. The developments of the mine, the mill, and the ship between them led to an elaboration of mechanics (pp. 395 f.) that was to be the basis of the greater revolution 200 years later and the inspiration of the intervening generation of seventeenth-century scientists.

The major transformations that characterize the great Industrial Revolution – from wood to coal as fuel; from wood to iron as material; from horse and water-power to steam-power; from single to multiple action in spinning – are all products of the ingenuity of workmen acting, as we have seen (p. 547), under the triple economic drive of increasing markets, the resultant shortages of traditional materials, and production bottle-necks due to lack of labour. They were made possible only by the availability of capital for making the new machines. All this might have happened without science, but it could not have been done so quickly.

370. Thomas Alva Edison (1847–1931), the inventor whose work in electricity, telegraphy, moving pictures and sound recording was to help lay the foundations for modern electrical and electronic industries. Here seen with his phonograph, invented in 1877: after a photograph by K. L. Dickson.

Indeed the very progress, interest, and profitability of the new machinery served to attract and generate science on its own account. The scientists became engineers, engineers learned science (pp. 591 f.).

The dominance of the practical man, the mechanic, and his employer, the captain of industry, lasted well into the nineteenth century. The precision metal-working on which the whole of our modern industry was based was built up at bench and lathe by common workmen on their own initiative (pp. 593 ff.). Only in the new fields of chemistry and electricity could the scientist, or more usually the enthusiastic amateur with a smattering of science, take the lead in devising new processes or new instruments.

The triumph of Edison marks the end of the era of the inventor and the beginning of a new one – that of directed scientific research in industry – which has gone from strength to strength in our own time. From now on the strands of industrial and scientific advance will be as closely mingled as they were before the dawn of civilization.

14.2 The Interaction of Scientific, Technical, and Economic Development

One conclusion clearly emerges from the parallel study of technical and scientific progress through the ages. It is that while science, as a conscious discipline, has existed since the dawn of civilization, it was not essential for any technical purpose until the sixteenth century, when it became indispensable for navigation; nor was it of use for many purposes until the nineteenth, when it was needed for chemistry and engineering.

Whether science in its early stages was really necessary for any purpose whatsoever is a more open question. In agriculture, a calendar, though useful, is not absolutely indispensable (pp. 121 f.) and the monumental architecture of pyramids and temples could well, one might have thought, have been dispensed with. However, that was not the view of the peoples of the time, for at least priests and rulers from China to Peru considered that proper and orderly contact with the heavenly powers was essential to economic and political well-being. Certainly we owe the science we have to that belief.

What enabled science, however, to get beyond a rather barren mathematical astronomy was the contacts it established with the techniques of mechanics, pneumatics, and ballistics in the Renaissance (p. 410), and with brewing, distilling, and textile chemistry in the eighteenth century (pp. 530 f.).

This brings out an important general conclusion, namely that while technical requirements often, though not always, provide the problems giving rise to new branches of science, scientific advances are always effectively *fixed* if they can be incorporated in practical and paying trades. The advance of optics was secured by the spectacle-makers (p. 425); that of magnetism by the compass-makers (pp. 434 f.). In our own times, jet engines and refrigeration look after heat theory; the radio industry takes charge of sound; the cinema industry of optics; not to mention the science of electricity that is part and parcel of the electrical industry.

The evidence from the dates and places of the bursts of scientific activity points to the need for rapid technical advance, and consequently for an economy favourable to effective capital investment, if science is to flourish. Though the financial demands of science are negligible in quantity compared to other expenses of production, money spent on science rarely provides any immediate prospects of return. Scientific expenditure is consequently in the nature of capital investment with a low rate of interest, but carrying the possibility of occasional big profits. The material factors holding up scientific advance are mainly due to its being starved of funds (pp. 1258, 1270), particularly where large-scale application was aimed at, as exemplified in the sad cases of Simon Sturtevant (pp. 415 f.) and Denis Papin (p. 577).

The satisfaction of the material needs of science is a necessary but not a sufficient condition for its advance; with it must go a sense of the excitement of novelty great enough to draw the most capable people into the adventure of science. On the other hand science may well be held back by the dominance of ideas belonging to an earlier period. One of the most persistent factors in holding back science has come from the very success of science itself. It is the belief in science as a means of attaining to absolute and permanent knowledge.

THE DOGMA OF SCIENTIFIC TRUTH

The study of its history has shown that science is not an entity that can be fixed once and for all by definition. It is a process to be studied and described, a human activity linked to all other human activities and continuously interacting with them. We have seen how it changes, not only by growing and extending into regions either absolutely new or hitherto left to simple common sense or myth, but also by continually working over its earlier findings with greater self-consciousness and greater power of integration.

This, however, is not a view of science that has always been held. In the past, and by some even today, the inner nature of science is held to be an autonomous system completely isolated from the social world. It is believed that there is an intrinsic and pure knowledge – a unique approximation to an absolute truth, to be achieved by a sure method and guarded by a passionate rejection of alternative ways of looking at things. The history of science is full of examples which show that the adoption of this attitude has been a sure way of arresting science, often while giving it the appearance of the greatest profundity and generality.

The supreme example is that of the Aristotelian–Averroist–Thomist synthesis (pp. 306 ff.), which provided a dominant world view and method of science from the fourth century before Christ till the seventeenth of our era. For over 2,000 years this was the pattern into which thought was frozen, and it took a violent change in the economic and political scene to shake it free again.

What we do not perhaps realize is that we ourselves are living through a stage which is just as much a transformation of science as it is a far more fundamental transformation of society. In science as in society, the efforts to save the situation by modifying, adding, and patching the old forms are likely to be futile. We will have to think out again, in the light of experience and controversy, the very bases of science itself.

THE RELATIONS OF SCIENCE AND SOCIETY

The relations of science and society are fully reciprocal. Just as transformations are produced inside science by social events, so, and in increasing measure, have social transformations been brought about through the effects of science. These are manifold, direct and indirect, operating on the material framework of society and on the ideas by which it is sustained and transformed. The direct effects of the material changes are the easiest to see, and are taken to be the main results of science by most people. They are obvious in the continuing and ever-accelerating stream of new techniques and improvements of old techniques in the transformed mechanical-electrical-chemical world of the twentieth century – a world inconceivable without science. The indirect effects of the material changes brought about by science are far more important. Though the first growth of science itself was a product of the economic and political factors, once science was established as a means of securing economic and political power its very progress became a factor in political and social life. No modern industrial state can exist at all without science; it cannot continue for long without making the fullest use of its resources of intelligence to advance science and to extend its utilization. The political patterns of our times are therefore increasingly a result of the material aspects of science.

But the material mode is not the only one by which science affects society. The ideas of science have a profound influence on all other forms of human thought and action, philosophical and political as well as religious and artistic. Here the influences are even more complex than on the material plane. The ideas of science are not the simple products of the logic of experimental methods; they are in the first place ideas derived from the social and intellectual background of previous times, transformed, and often only very partly transformed, by passing through the test of scientific experiment. So it was, for instance, with the idea of the natural law of Newton, which was a reflection of the establishment of a legal rather than an authoritarian form of government (pp. 488 f.), or with the theory of natural selection and the struggle for existence of Darwinian evolution, which was a reflection of the free competition of the full capitalist era (p. 662). But though these ideas

371 a, b. The earliest use of the magnet seems to have been by the Chinese, who knew of its north and south seeking properties. The illustration shows a magnetic chariot, carrying a figure that always points towards the south. Magnetism is now used in the most sophisticated research techniques in nuclear physics, specially prepared and accurately machined magnets being used to confine electrically charged nuclear particles within paths desired by the experimenter. This photograph, which shows a great contrast to the old Chinese drawing, depicts the flight tube for particles that runs from the linear injector to the main ring of the synchrotron. Some of the magnet units can be seen in position on the right of the picture. Photograph taken at Meyrin, Switzerland, in the research laboratories of CERN (Conseil Européen pour la Recherche Nucléaire), which was formed in 1955.

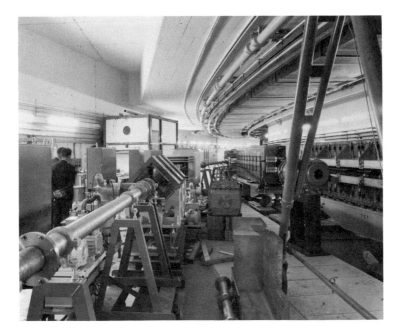

emanate from earlier social forms, once taken into and used by science they in turn are both reinforced and transformed. They are used to give scientific sanction to social practice, as has been shown in many examples in its history, from the Aristotelian world-picture onwards (pp. 207 f.).

The full understanding of the ever-changing relations of science and society involves taking into account both the material and the ideological factors. No society can do without the amount of science required for the working productive forces. This amount, however, as we have seen, has been very little until recent times. Whatever science a particular society possessed in addition to this minimum was that needed, in conjunction with philosophy and religion, to buttress the current social system. Incidentally, it provided an intellectual occupation for those few members of the ruling classes who had the inclination to use their leisure in this way. Such science could not be expected, in itself, to be a stimulus to technical change or to determine the pattern of production, but in the presence of other factors it could be invoked for that end. These were for the most part economic. The machinery of the Industrial Revolution was not the simple gift of inventors – there had been ingenious men in plenty in earlier times – but, as we have seen (p. 520), it grew in response to the availability of capital and labour and to the opportunities the market offered for profit. Once in existence, however, a new process or machine – or, better still, a new scientific principle capable of generating many new processes and machines, like Faraday's magnetic production of electric current – must necessarily change the condition of manufacture and give new opportunities for economic changes.

The sciences, social as well as natural, have appeared throughout history in this catalytic role as agents, though not generators, of social change, and have in the process been changed themselves. The track of scientific and that of economic development run so close together in place and time that their association cannot be a chance one. Something of the nature of that link between scientific and economic activity has been brought out in these pages; but there is not, and indeed there cannot be, any simple formula relating them. It is rather, as has been explained, that the economic and political conditions of particular places and periods are especially favourable to the growth of science, in setting problems and in rewarding successful solutions. The discovery of problems is in fact more important than the discovery of solutions: the latter can be arrived at by experiments and logical argument; the former only by imagination stimulated by experience of difficulty.

THE ROLE OF GENIUS

It is sometimes argued that this economic connexion touches only unimportant parts of science and that the great discoveries are made by men of genius unaffected by conditions of time and space.[8.7; 8.53] Copernicus, Descartes, Harvey, and Linnaeus may be cited as examples of men who worked far from the centres of technical advance of their time. This is only partly true, for the sixteenth century in Poland, the seventeenth in England and France, and the eighteenth in Sweden were all periods of national expansion. Yet it remains a fact that at least in the first and last cases the countries they lived in were not central, and that neither Harvey nor Descartes had close industrial connexions. The reason for this apparent discrepancy is that it is not where a scientist is born or even where he works and dies that matters. Once he has found his life work he is fully mobile, he can work wherever he can live. What matters is the centre of the tradition that formed him. For Copernicus and Harvey it was Italy, which had only just passed the peak of its economic and cultural greatness. For Descartes it was the cultivated world of French society just about to enter the *Grand Siècle* (pp. 441 f.). Linnaeus is the exception; largely self-taught, he conquered the kingdom of the plants by his almost religious faith, single-mindedness, and hard work (pp. 636 f.). But his system would have died with him if there had not been in his time a world of gardeners and botanical collectors eager for any means of setting their herbaria in order. The seed of science is not sown nor does it grow unless the social ground is well prepared for it by economic activity (p. 45).

The progressive growth of science comes from its continually renewed interconnexion with industry. As human society develops, the part that technique and science play in it is, as we have seen, an ever-increasing one. So also is the part that conscious and logical science takes in technique. It is difficult for us to discover amid a mass of ritual and traditional techniques of ancient times any implied logic that we can call science (p. 77). We can recognize it as such only because we know what will come later. Today science appears as an institution in its own right, with its own tradition and discipline, its own specialized workers, and its own funds. What is more important, every aspect of life – industrial, agricultural, medical, administrative, and above all military – increasingly needs the help of organized science for its day-to-day operations, and is entirely dependent on science for its progressive development. This trend, which shows no sign of abating but rather of accelerating, lies along the line of the increasing consciousness of human activities, carrying with it greater control over environment through the understanding of its laws.

THE EMERGENCE OF SCIENCE
AS A MAJOR FACTOR IN SOCIAL ADVANCE

The emergence of science as an important agent in the social sphere is a critical and irreversible step in general human history. Together with the economic and political changes with which it is inevitably coupled, it is an event of the same order of importance as was the emergence of the human race itself or of its first civilization. Such crucial changes have negative as well as positive effects. Not only do they make new things possible, but they also make old ones impossible. Once there has been a break-through on to a new level of complexity, the possibility of similar jumps occurring elsewhere rapidly vanishes. Once, for instance, the green plants had spread on to the land, there was no more room for plant-forms with any other metabolic basis to do the same.

The growth of modern science, linked as it was at the outset with capitalism and the Industrial Revolution, has been pervasive and irreversible. The new way of saying things and thinking about things which arose from a study of the techniques and arts has, by transforming those techniques, led to a much more comprehensive approach to human as well as to technical problems. By doing so it has also prevented, and will prevent, any solution that rejects science from being ultimately effective.

This is well brought out by an instance from very recent history. The Nazis were from the beginning violently anti-rationalist, and in favour of a mystical system which alone would justify them, while it served to conceal their real ends in supporting monopoly capitalism. Naturally, therefore, they were opposed to science, but nevertheless they could not do without it because they needed science for their one effective means of action – war. Any State or class which does not, or cannot, use science and develop it to the full is doomed in the present world to decay and destruction (p. 1153).

The transformation of economy through the use of science has been a very recent event. It can be said to have reached critical importance only at the beginning of this century, and its subsequent development has been so rapid that it has been impossible to allow science to seep slowly, as it did in earlier times, into the existing forms of economy. It has appeared as a new agency, one that is changing things rapidly in relation to the life-span of men. In the previous transformations of civilization the process of change was much slower: new ways came in with new generations, and where there has been open struggle it has been because certain social groups or certain peoples remained untouched by the new forms and resisted them.

In this age the new capacity for handling the material world has been gained well in advance of the appropriate patterns of cultural, political, or economic life. This situation is often described by the pundits of science and religion in terms of man's material powers having outrun his moral stature, with the implication that science must be halted or turned back until man has been spiritually regenerated. This, however, for reasons just given, is not in the least likely to occur; science is too useful, if only for destruction. Rather must we look for the opposite solution and seek, through a better-ordered society, to raise the moral level of humanity. Though struggle and difficulty are implicit in such an attempt, it has every hope of rapid success. For with science has also come the ability, not available in earlier times, to understand social situations and also, through conscious social organization, the means of solving the practical problems of the transition. The value of the study of the place of science in history lies in the understanding it gives to the economic and technical transformations leading up to that in which we are now all involved.

14.3 The Track of Scientific Advance

One major historical question still remains to be discussed. How can the study of science in history help to explain the particular track of industrial and scientific advance? How was it that it followed the course it did: from the fertile crescent of Babylonia and Egypt, through the Mediterranean from Greece to Italy, to settle in its greatest formative period of the seventeenth century round the North Sea, from there to spread in our own time all over the world? It is evident from what has been written here, notably in the introductions to the first four parts and in this chapter, that here is no case of rigid predetermination, though clearly certain conditions must be satisfied before any area can become a centre of civilization.

Within limits, no region can be a centre of economic or cultural advance for long without having adequate natural resources – good hunting grounds, cornland and timber, coal-fields and iron ore, oil and

372. It is clearly of importance to increase drinking water supplies as the world population increases. De-salination of sea water may be carried out by distillation. A salt-water conversion plant at Freeport, Texas, located in a coastal marshland where it can draw water from the Gulf of Mexico.

hydro-electric power – according to the stage of technical development. An equally essential negative condition is the absence of a climate leading to endemic disease or to general debilitation, such, for instance, as ended early Maya civilization. These are only permissive conditions, for at each stage not one but many regions have had the necessary resources and climate. Much of the world – covered by tundra, deserts, tropical rain forests – has been ruled out until our own time; but there is plenty left.

Which of the geographically possible areas will become a focus of advance depends rather on the forms of society – the productive relationships – and their economic and political concomitants. Here geographical factors may still play some part; complete isolation may hold back development, as happened in the case of the native American civilizations; but on the other hand too easy access from other regions, as in the case of Babylonia and south-eastern Europe, may lead to regression in the face of the penetration of less civilized tribesmen. More decisive, especially in later times, has been the persistence of an internal drive to change, provided by the sequence of class struggles –

373. Technical advance may well lead to new living conditions and to new means of food production for an ever expanding world population. Underwater farming and living is now under experimental investigation, while underwater oil prospecting is already an accepted technique. Geologists collecting samples.

technique and culture being the weapons for securing the mastery of each class in turn, and being advanced in the process.

In the earlier ages of the world it would seem as if the evolution of first a village and then a city culture could and did occur wherever material circumstances were favourable (pp. 98 f.). Given the considerable geographic insulation of such cultural centres, this did result in the appearance of some half-dozen patterns of civilization, including the isolated civilizations of Central and South America. Once each was established it preserved an internal traditional continuity in material techniques and social forms which made it, within limits, resistant to outside influences. Each area – the Hellenic, the Indian, the Chinese, the African, the Amerindian – acquired distinctive features, ranging from basic ideas to styles of ornament.

To admit the existence of different patterns of civilization does not imply accepting the mystique of treating a civilization as a biological species which evolves, spreads, or becomes extinct, or, even more extravagantly, as a spiritual being that is born and dies, as in the recently fashionable beliefs furthered by Toynbee and Spengler.[8.63; 8.60] A civilization is a convenient way of grouping together developments within a given area, sometimes a very wide one indeed, that depend on one continuous intellectual and technical tradition. Looked at closely, civilizations are elusive and indefinite, continually changing and merging into each other. How far can one say that the Chinese Buddhist culture belongs to Indian civilization? The central and enduring features are the means of production, the pattern of agriculture and industry, and the productive relations, whether classical, feudal, or capitalist. Language and literacy and mythical tradition may help to ensure coherence, as Homer did with the Greeks, or the Bible with the Hebrews.

The general pattern of agriculture and industry is one of relatively immovable forms and institutions; but much of civilization can and does spread easily. Thanks to the traders and itinerant craftsmen, technical devices, which can be made anywhere once the trick is known, such as the wheeled chariot and the still, were spread far and wide. The technical level which all these agricultural civilizations reached was very similar. Thanks to the wandering scholars, the useful intellectual ideas, the sciences of mathematics, astronomy, medicine, and a little alchemy, were also carried almost everywhere, and tended to form a common pool of lore interpreted differently to fit the prevailing traditional and religious ideas. As we have seen, until the fifteenth century at least, the relations between the major civilizations were on a fair give-and-take basis. There was no marked superiority on the surface, and an intelligent Italian like Marco Polo, though he found much to wonder at in the urbanity and culture of the Chinese, had no difficulty in filling a post in their administrative system.

Yet only one of these civilizations was to give birth to the next phase of advance – economically with capitalism, technically with science. In material equipment there was little to choose between the main centres of civilization in the Middle Ages, nor was there much intellectually. Thanks to the middleman role of Islamic culture, the joint heritage of Hellenic, Mesopotamian, and Indian cultures in mathematics and astronomy, which was to serve as a basis for further advances, was sufficiently widely known to enable that advance to be possible anywhere. I would suggest two decisive factors that may have given Europe the advantage: positively, the specific incentive to industry provided by a growing market and rising prices in the sixteenth century in a relatively underpopulated area (pp. 411 f.); and negatively, the absence of heavy traditional blocking, such as that brought about by religious bigotry in Islam and India (p. 273), and by bureaucratic obstruction in China (p. 312).

These conditions effectively limited the origin of capitalism and the breakdown of feudalism to Europe, and in Europe to certain favoured regions. What were, in sum, the special advantages of Italy, France, and the North Sea countries in the fifteenth and sixteenth centuries? Italy and France had been the richest and most populous areas of the Western Roman Empire, they retained most of the old traditions, especially those of city life and craftsmanship. However, their basic natural resources could not compare with those of the countries round the North Sea, including Flanders, Holland, Friesland, and the lower Rhineland, as well as England (pp. 286 f.). Once the forests could be cleared and the land roughly drained they provided some of the best agricultural land, with the most reliable rainfall in the world.[8.61] They still give the record yields per acre. England produced the world's best wool, which was to be the basis of her greatness.[8.54] The North Sea provided abundant supplies of fish. Indeed the countries round it lacked little but salt, dyes, and spices. At first they had plenty of timber; as that gave out they found a substitute in their coal outcrops and learned to make and use more iron. Further, over most of the area water transport ensured that materials and goods could be readily interchanged. These are all necessary conditions; no area much less well endowed could have carried, through successive technical transformations, the burden of an expanding culture, which is always apt to

374. The recent finds of oil and natural gas under the bed of the North Sea is an example of the use of technology for improving fuel resources. Such developments may have a profound effect not only on the economy but also on political alliances. The drilling platform *Sea Quest* on location in the North Sea.

run up to the margin of available resources before new resources can be mobilized.

All these favourable conditions would have been of no use unless the social means had been evolved to exploit their advantages. Under the feudal system this could not be done; but under capitalism it was possible to make use of the resources with ever-increasing efficiency as the profits of earlier successes were ploughed back into productive enterprises. Further, capitalism could use improved techniques and encourage science to develop radically new ones.

It was not easy to secure a base for capitalism. The first attempt in fourteenth-century Italy had collapsed by the sixteenth, under the combined effects of its own inner tendencies to relapse into feudalism and the attack by the feudal forces of Italy and Spain (p. 329). The establishment of the first stable base in the Low Countries in the sixteenth century was not achieved without violent and even desperate struggles (p. 411). Once established, however, the enormous economic advantages it offered ensured its spread, first by example and then by conquest, all over the world. By the mid seventeenth century it was clear that no country that had not adopted capitalism could hold its own against those that had.

Further, the stimulus early capitalism gave to science ensured a crushing advantage for the new science over anything that might come from intellectual developments in older cultural centres. In fact neither in India nor China, despite their earlier cultural achievements, was there any significant advance after the sixteenth century until our own times.* Any later flowering was prevented by the economic and political encroachments of the foreign capitalist powers. From the seventeenth to the early twentieth century, science and intellectual culture became a European monopoly more and more concentrated in the centres of heavy industry. Today we are witnessing the beginning of the reverse process, starting first in Japan (p. 715): a widespread decentralization of science and technique, which, despite every effort to keep it in the hand of the masters of 'Western Christian Civilization', will show that its advantages can be reaped by men of all colours and traditions. It is in this present world, with its conflicts between the restrictive and liberalizing forces, rather than in any imagined ideal and timeless state, that we have to see the problems of science in relation to society.

14.4 Science in a Class-divided Society

One other aspect of the interaction between science and society remains to be discussed, the effects on science of the class divisions in society which have been in existence since the very beginning of civilization. Brought up, as most of us have been, in a class-divided society, and so taking it for granted, it is not easy for us to see what this implies for science. In fact, as this book should show, the influence of class divisions has since their inception permeated science both materially and ideologically, and has affected its structure, development, and use (pp. 49 f.).

The successive transformations in technique that have occurred from the beginning of civilization till the present day have been motivated at every turn by the interests of individuals and groups of the ruling class of the time. Any benefits they have brought to other classes have been incidental. Slaves and serfs, if they are to do their work and provide food, must be kept alive. As the machinery of civilization becomes more complicated, some of the workers may have to receive appropriate education.

We may rightly admire the Pyramids as a feat of architecture and engineering, but what they do represent is a waste of the labour of tens of thousands for the hypothetical benefit of the soul of the Pharaoh, and for the real prestige and rake-off of the priestly contractors. In our grandfathers' time, we are reminded, by the only too permanent relics they have left us in ugly towns and blasted countryside, of the price the people had to pay for the progress of the Industrial Revolution. And now, in this age of science, how terribly little of the new knowledge, skill, and ingenuity is being used for any bettering of human conditions, and how much for sheer destruction!

As a historic fact we owe the development of science, as of other aspects of civilization, to the operation of class societies. It would be pointless to consider how else science could have grown, but stupid to assume that, for no other reason, it must continue under the same auspices. Class societies have left us some very fine things, but very bad ways of getting and using them.

375. The application of fundamental science to technology has brought both new materials and new means for their testing and for quality control. Here a quality control laboratory is testing high density polyethylene to be used by manufacturers of plastic products.

It has been made clear by many of the examples in earlier chapters (pp. 553 f.) that it has been the operations of class interests that have again and again held up the advance of science. Successful applications in war or profitable applications in peace have been the only criteria for technical advances. In turn the establishment of new techniques has been the only way available to fix the achievements of advancing science. We have seen (pp. 624, 547 f.), notably in the histories of chemistry and electricity, that it is only where a new chemical or piece of apparatus appears in quantity as an article of commerce, often many years after its first discovery, that it can become the basis of further scientific advance.

Considering the available technical skill and intellectual capacity at different periods, it is apparent that these were rarely if ever the major limiting factors in industrial progress. Even in the progressive nineteenth century, it was lack of anticipated profit that kept short-sighted and tradition-bound capitalists from embarking on new enterprises long after they were technically feasible (pp. 612 f.). Sooner or later one of them would make the jump and then all the others would rush in. The net effect has been that the advance of science has been far more delayed and erratic than it need have been.

Now that the process of haphazard scientific application has given way to a system of organized industrial research, it might be thought that these delays would be a thing of the past. Actually, if the obstacle of lack of capital has lessened with its greater concentration, so also has the incentive that competition used to offer; and the great monopolies that have replaced the multitude of competing firms show themselves in no hurry to make radical innovations. Examples of this have been given in earlier chapters (pp. 781, 822). I need only cite one in the very centre of a scientific industry. The principles underlying fluorescent lighting were known even before filament bulbs came in; but forty years had to pass before adequately supported research made it commercially possible. The well-advertised successes of monopoly-directed science, such as nylon and television, are apt to blind us to the things that are not done (pp. 828 f.). For the possibilities of applications of science are known only to a few. But these see well enough how little of the immense resources of actual knowledge is used, owing to the misdirection and restriction of everything that does not promise immediate profit. In one way or another, science has been kept from the service of mankind. It is not the particular forms of class societies that restrict science but their very essence: the exploitation of man by man.

The existence of class-divided societies does not affect only the material consequences of knowledge, it cuts deep into its roots in ideas. The literate and cultured are the ruling class, and the basic ideas that find expression in literature and science are inevitably tinged with ruling-class preconceptions and self-justifications. At the same time the fund of practical experience which comes with the daily work that maintains all society is cut off from literary expression and from academic knowledge. It is significant that the periods of greatest productivity in the arts and sciences – the time of the early Greeks, the Renaissance, the Enlightenment – have all been those in which for a short time class barriers were partially broken down. In such periods it is the aim of a rising class to capture literacy, culture, and science and in the process to make them far more generally available (pp. 1151 f.).

It is in its basic philosophy that class influences have most affected science – the natural sciences, within the limits imposed by experience; the social sciences, entirely. For many ages, as we have seen, the two were not separate. The aim of a comprehensive philosophy, as it seemed to the Greeks, was to render a coherent account of the whole universe, using the order which it revealed to justify, in particular, the social order of the city state. The Platonic ideal, just as much as the Aristotelian mean, were politico-scientific constructs (pp. 179 f., 193 f.).

When reason failed, faith was called to its aid, if not to justify then at least to make tolerable, as a step to heaven, the thoroughly vicious and unjust social system of the Roman Empire. Science retreated; when it advanced again it remained for long shackled with a scholasticism the main aim of which was again to justify from scripture and reason the crazy inefficiencies and gross injustices of the feudal system.

The decisive step, which was taken at the time of the Renaissance, was to separate the spiritual and material worlds, leaving a natural science free enough to turn to practical profit, but with ideas still drawn from the old theological-philosophical system. However great the material successes of the new science, these ideas have remained deeply submerged during the advancing optimistic phases of the Industrial Revolution, and are coming to the surface again as the system decays (pp. 1157 f.). As in former times, the need for a philosophy justifying an intrinsically unjustifiable system of class rule is producing an idealistic distortion of philosophy. Some of it is a naïve identification of the superior, the spiritual, the ideal, as the special apanages of the upper classes, and insistence on higher things is used to cover the order of society with divine sanction (pp. 132 f.). In a more sophisticated form, adopted particularly in Christianity, spirituality was democratized on condition that power and wealth were left, in this secular vale of tears, to the temporal authorities – the 'pie-in-the-sky' philosophy, which can so conveniently substitute charity for social justice (p. 256). In the more extreme form of mysticism, retreat from an unreal world is advocated, with the result that it becomes quite irrelevant how it is run. These semi-religious approaches are a means of permeating the underlying fabric of science itself and persistently drawing it away from the real world.

SCIENCE AS A PRODUCTIVE FORCE

These considerations of the class character of science are not merely historic and academic; they should help us to understand the conflicts that are dividing science as well as society in our own time. Once both the scientists and the people at large grasp the social significance of science, they can no longer go on in the old way, and let science and technique grow up casually under the impulses of restrictive and distorting private interests. The new knowledge of the nature and powers of science cannot be thrust aside, but acceptance of this means accepting a social responsibility for the fostering and direction of all science (pp. 39 f.).

For the same reason it is useless to repine at the way science has grown in the past. Science, like all other human institutions – language, arts, religion, law, and politics – has achieved a content and a power which transcend the means or the motives that helped step by step to build it up. With practical technique, with which it is associated ever more closely, natural science has a greater independence than the other more purely social institutions, because it is securely anchored in the material world, in the properties of animate and inanimate things.

Of even greater importance is the fact that science is becoming in recent times what technique has always been, an indispensable part of the productive forces of society. The technical *know how* now needs to be backed everywhere by the scientific *know why*, in order to maintain the life and growth of a modern community. Science may in part have been an ill-gotten and unfairly hoarded treasure, but it is a treasure none the less. Now it must be for all to spend and to increase.

14.5 Science in the World Today

Before we can usefully discuss the bearing of historical studies on our current problems it is first necessary to look more closely into the present world situation of science and to relate it with the actual distribution of political and economic power, condensing into a few paragraphs and supplementing the information given in Part 6.

The balance of science between the different regions of the world is a grossly uneven one, but it is also changing with great rapidity. Its distribution corresponds, for reasons already given (pp. 706 f.), very closely with that of heavy industry. More than nine-tenths of the scientific manpower of the world is concentrated within a few score miles of some dozen major coalfields and of about the same number of capital and port cities in other areas. The population of this rapidly growing industrial sector of the world is some 730 million, or about 23 per cent of the whole world population. The rest of the world is agricultural, of which the greatest concentrations of population occur in the basins

376. Heavy industry is needed in countries previously dependent on other powers. A steel works in India, at Bhilai.

of half a dozen great rivers and highly cultivated islands like Japan and Java, occupying a twentieth of the habitable land surface and carrying a total population of some 1,400 million peasants, so far very badly served by science. The remainder, apart from virtually uninhabited desert, mountain, and tundra, carries a population of 1,200 million, also mostly peasants – one-third of the world population – though occupying 90 per cent of the habitable surface of the globe. (Map 5, p. 1314).

These purely geographical divisions acquire a meaning for present and future science only in the light of the political and economic systems under which they are now being administered. Today the world divides most naturally into three sectors which may be called the capitalist, socialist, and former colonial sectors, now rapidly becoming a neutral bloc of states no longer part of the 'Free World'.

First there come the highly industrialized imperial states, new and old: the United States, Britain, Germany, France, Italy and Japan, together with the smaller and weaker industrial states of Europe and America which are economically, strategically, and politically dominated by the United States. These between them comprise a population of 600 million of whom about 420 million are industrial and 180 million agricultural. The hard centre of capitalism is even more restricted, effectively to the eastern and Great Lakes States in America, to industrial Britain, to N.E. Europe and to Japan, for these between them produce nearly two-thirds of the world's steel, the key material of modern industry. Here, between crises, industrialization and production increase rapidly, but the pace has been so much greater over the whole period in America than elsewhere that production is becoming ever more concentrated there. It is true that in the European countries for the last decade there has been an astonishing capitalist growth, but this is closely linked with the capitalism of America. Europe is, indeed, the chief field of investment of American capital. The real control of this industrial complex is by no means unified, for rivalries persist and new combinations of forces are continually being made. But the controlling oligarchy, the active directors of the fifty great financial and industrial trusts of the world, are a small and exclusive band, probably not more than 100 persons in all, of which the dominating interests are in the United States.[8.49]

The second sector of the world consists of those countries that have at various periods since 1917 thrown off capitalist domination and have reached different stages on the road to a classless communism. The total population of this group is some 1,100 million, or a third of the total world population. Its present degree of industrialization is low, 230 million or 20 per cent, 30 per cent if China with its predominantly peasant population is excluded. The significant character of this sector, in relation to its use of science, is that it is undergoing an extremely rapid industrialization and that, in contrast to the centralizing tendency of capitalism, this is being done in a widespread manner so as to raise the standard of industrial production equally in all areas, to develop agriculture in industrial areas and industries in the agricultural areas. This also implies the active and planned use of science in both industry and agriculture.

The third sector of the world comprises the remaining areas of Europe and Asia, and the whole of Africa, Oceania, and Central and

377. In many African and Asian countries, great numbers of people are still living in sub-standard accommodation. In the slums of Hong Kong more than 300,000 people are housed in shanty conditions and UNICEF has to provide medical supplies. Photographed for UNICEF by Jack Ling.

South America. This is predominantly the area of raw material and food producers for the old imperial countries. Of the total population, 1,400 million, or nearly half of the world's peoples, only 8 per cent are industrial. The rest, with the exception of the privileged farmers of the British Dominions, are peasants, plantation workers, or serfs, with a very low standard of living.

This is shown in the most elementary way in the absolute consumption of food. The peoples with a consumption of over 2,700 calories per head per day are all either in the first sector or in such countries as Australia and New Zealand. At the other extreme, in most Asian and African countries the consumption is under 2,200.[8.68] This physical misery produces greater susceptibility to disease and there are fewer means of combating it. The average expectation of life in India is only 32, as against 70 years in Great Britain.[8.70] With it also go mass illiteracy and a poverty-struck agriculture that yields ever lower returns.

Of this section, 20 million people, or less than 2 per cent, are directly under the imperial rule of the old industrial powers, and another 15 million – the black population of South Africa and Southern Rhodesia – are under the domination of similar interests in South Africa. Another 820 million, or nearly 60 per cent, including by far the most populous countries of India, Pakistan, and Indonesia, have achieved their political independence since the Second World War, but are economically dominated by imperialist powers new and old. This expresses itself in holding back industrial development in the interest of the imperial powers and concentrating the whole economies of the countries for the production of agricultural and mineral raw materials – strategic metals and oil – which can be profitably extracted by cheap labour. The soil is exhausted by single-crop agriculture, often on foreign-owned plantations, and the products are processed only just so far that they can be removed most conveniently. Finally, the profits arising from these enterprises leave the country and cannot be used to build it up economically.

The lack of industries in this colonial or semi-colonial section of the 'free' world ensures also that they should be governed most undemocratically in the interests of the great powers, either directly by foreign officials or indirectly by the nominees of local landowners or business men, though their control is rapidly, under popular pressure, passing out of their hands. Needless to say, such conditions find little place for science, though there is a growing awareness of its importance. There are fewer scientists in Africa and South America combined than in Holland.

The two most characteristic features of scientific research and development in the capitalist world today, and particularly in the United States, are those of concentration and militarization. At no previous period in history have industrial production and, to an even higher degree, scientific research been concentrated in such a small part of the world, and never has the proportion of military to civil research been as great as it is now (p. 847). Both these features are consequences of the development of monopoly-controlled industry.

The pursuit of the maximum profit (pp. 705 f.) is here the predominant factor in determining the balance of effort between industries and between the sciences that serve them. Throughout the whole history of capitalism technology has been developed, and science has been called on to help only when it seemed to be the most profitable way of organizing production. Where, as for example in the British textile industry in the last hundred years, there was cheap labour and old machinery the value of which had been written off, no serious effort was made to improve technique or use science.[8.40-41]

Effectively, the employment of scientific research is a form of capital investment. The recognition of this relationship has been a very recent one. It was first seriously discussed by Bichowsky in 1947 and is now generally accepted.[8.15] Only where new capital can be invested is it worthwhile even thinking of undertaking research. Even then, until very recently, the amount spent on research and development represented only 12 per cent in Britain and 17 per cent in the United States of new capital investment.[8.38] With increasing monopoly, what was an automatic, unconscious process has become deliberate policy. Calculations of profit on an unprecedented scale – up to 50 per cent in America, where the majority of companies consider that equipment should pay back its cost in between two and five years – are a necessary precondition to serious capital investment, and for the development and research that would make it possible.*

The monopoly control of science is so well covered by the techniques of advertising that the public is made to believe that the sums spent on industrial research are primarily for its benefit, instead of their being channelled, as they are even in the civil section, to produce goods such as television sets and fancy drugs, where the margin of profit is greatest.

It is the same demand for maximum profit that has given, in recent years, the heavy bias of technology and science towards military uses. Profits there are enormous: the public pays without asking awkward questions, and the resulting goods do not clog the market. They can be expended in wars or, if that fails, scrapped in a few years as obsolete. The demand for them is also reinforced by every means of propaganda needed to keep up war fever and justify military expenditure. One consequence of this has been the militarization of science, which has already been discussed (pp. 831 f.), and will be referred to again (pp. 1257 ff.), with all its consequences of secrecy, screening, and witch-hunting.

In one way or another, directly or through government agencies, science in the capitalist sector of the world has come under the control of the small number of big monopoly firms. In the United States,[8.1] the universities are already in their hands; their representatives sit on governing bodies; they provide the funds or arrange government grants; they give employment to the graduates; they can make or break leading scientists; their influence is predominant in the scientific societies, which can continue to exist only thanks to their subventions. Only the advantage of maintaining, in the eyes of a well-meaning public, the appearance of academic freedom and of their own benevolence in upholding it, prevents them taking the whole show over openly.

Indeed ever since the beginning of the century the policy of supporting research has been, along with philanthropy and the patronage of the arts, one of the deliberately adopted props of the moral position of the great monopolist dynasties of Rockefeller, Mellon, Ford, and Du Pont. For an expenditure of a minute fraction of the profits gleaned from many years of exploitation they have been able to pose as the main patrons of disinterested research. Since the Second World War their place has been taken by the government, which has become, as already described, the main source not only of university research funds but also of industrial research. This new arrangement, which operates in Britain as well as America, is that of the research and development contract, almost exclusively for war material. It has proved very convenient for monopoly firms, as the government pays the costs and takes all the risks, while, once in production, industry takes all the profits. In 1961 the US Government paid for 64 per cent of electrical research, 46 per cent of instrument research, and no less than 89 per cent of aircraft research. These between them account for about $6,700 million or over 60 per cent of the whole industrial research expenditure of the country. In Britain comparative figures are harder to find, but the actual distribution of research expenditure among a few big firms in the aircraft

378. Research into public problems such as air pollution is undertaken by government sponsorship. Air pollution studies being conducted in a wind tunnel at the College of Engineering, New York University. Photograph by Irwin Gooen.

and engineering industries shows that much the same process is going on.[8.24–5]

THE ORGANIZATION OF RESEARCH

It is against this background of increasing industrial and governmental research that we must look at the present organization of research in capitalist countries. It has now become something very different from that which started with the academies of the seventeenth century, though these remain as honorific bodies. Its purpose has changed and its scale has become enormously greater. Its concern is with the use of science both in the operation and in the evolution of the economy and administration of states whose major technical interest has become the preparation of war. The scientific organization of today is not limited to, and is in fact not mainly concerned with, the internal development of science. Nevertheless, as the very existence of a vastly more costly body of scientific research has come to be almost entirely dependent on state and industrial finance, the future of science is bound to be directly and profoundly affected by the way that research is organized.

By the end of the nineteenth century the older form of scientific advance, through the activities of individual scientific men, either of independent means or earning their money as private consultants, had virtually ceased (p. 708). Instead, the main advances in fundamental science were concentrated in the universities, where the newer functions of research were added to the older ones of teaching. This form has since become an almost general one, the only exception being the existence of a certain small number of research foundations, but even these tend to be attached to the universities. The advance of science appeared in the first place as a by-product of general teaching, but as its importance grew the tendency was for research to dominate teaching and for science teaching itself to become limited to an introduction to research.

Already, however, the researches for which the students are trained lie largely outside the universities, in industry or government service. Apart from its beginnings in the electrical industry at the end of the nineteenth century, the major part of strictly industrial research dates from the second decade of the twentieth century. But the growth of industrial research has been of a rapidity quite outstripping that of the older forms.[8.44] It is probable that between 1920 and 1960 industrial

379. The wind tunnel has proved useful in testing the strength of structures. A model of a suspended section of a suspension bridge.

research throughout the capitalist world increased by a hundredfold, and already the vast majority of a greatly enlarged class of active scientific workers are employed by industry or in the industrial sections of war preparations. Originally the purpose of industrial research was to apply the results of science to the needs of production. As time went on industrial research bodies tended to accumulate more and more basic knowledge, especially in physics and chemistry, and to draw into their service able research workers in fundamental science. As a result the centre of gravity of science is moving more and more into the industrial sphere, with many bad results, not only through the spread of secrecy but also through the removal of any control of the general direction of research from the body of competent and independent scientists.

GOVERNMENT AND MILITARY RESEARCH

The most recent development in the organization of science, however, has been due to the large-scale intervention of governments. It is true that ever since the seventeenth century some of the subvention of science has come from government sources, but it has been almost entirely concerned with the special services of astronomy or cartography or of the proper standardization of weights and measures. In capitalist countries there was in fact, until these last few years, a definite and strong objection to the intervention of government in science, because this would interfere with the proper competitive utilization of sciences for the profits of individuals and corporations.

That objection has been, as we have seen, entirely overruled because of the new common interest which government and monopoly firms have found in war research. The process took time: in the First World War science, neglected at the outset, became before the end a minor though essential auxiliary for the production and servicing of new devices such as the aeroplane or wireless; in the Second World War science was important from the start and by the end became a dominant factor, not only in the evolution of new weapons such as long-range rockets and the atom bomb, but also in the co-ordination and direction of military operations themselves.[8.11; 8.18] During the war practically the whole of science in Britain and America was turned to war service (p. 843).

Even after the war the subvention to science by governments for the preparation of new and ever more scientific wars continued to be multiplied by large factors. Thus in England, the Parliamentary expenditure on science rose from under £5 million in 1937 to some £78 million in 1947 and to £385 million in 1962; while in the United States it rose from fifty million dollars in 1940 to more than 600 million in 1945 and reached 16,000 million dollars in 1963. The increases of expenditure, both on industrial and government science, do not imply corresponding increases in the number of scientists, though this was great enough. The number of qualified scientists in Britain in the Government scientific service rose from 743 in 1930 to 7,059 in 1962, a nearly tenfold increase. Nor does this imply a corresponding increase in the quality of the new knowledge produced – rather the reverse. The greatest amount of expenditure is on costly apparatus and equipment and on the very large number of secondary staff. The growth has been so rapid that it has definitely hindered the advance of fundamental and basic industrial science for civil purposes. In Britain, indeed, there was a definite setback in 1950. It was so sharp as to draw a protest from the council of the Department of Scientific and Industrial Research itself:

> Basic research is hardly worth doing if the effort that can be devoted to it is insufficient to secure steady progress, and the suggestions made from time to time to reduce the present meagre effort appear to us to be unfounded.[8.28]

The relative backward condition of British civil science has at last been officially recognized. Pending the election of 1964, both major parties promised to give great prominence to civil scientific research. It is doubtful, however, whether they will be able to do so as long as such a large proportion of effort is given to military research.

This very increase in the material needs of science tends to make the contribution of government to science absolutely dominant, and for capitalist governments this is given primarily for military ends. In 1962, 64 per cent of government expenditure on science in Britain was for military purposes, involving 52 per cent of government scientific personnel. In the USA the corresponding percentage for expenditure was 90. This influence does not stop at the level of application; it penetrates the whole of research. In the United States, the Department of Defense and Atomic Energy Commission finance about 25 per cent of the country's basic research.[8.8] The supply of scientific workers, largely for war preparations and eventual war, has become a subject of anxiety, and accordingly governments have largely taken over the finance of the universities. In Britain, for example, the government grant to universities has increased eightfold since the war and now represents some 70 per cent of their income. Despite this there remains a chronic shortage of trained scientific workers both in Britain and America,[8.21; 8.36; 8.37; 8.39; 8.62] the cause of which is to be found in the limitations which the class system puts on education (pp. 1148 ff.). What

380. A view of Essex University, which was founded in 1964. It has some liberal cross-fertilization of studies and no overt distinction between students and staff. The architectural design is unusual, with a contrast between living quarters and university buildings for work. The photograph shows (left) a general university building, partly administrative, partly science faculty; (right) the library; and (centre) three halls of residence, each of fourteen floors and with separate personal apartments.

might be a break-through out of this paradox is now being provided in Britain by the drive to widen the basis of higher education, particularly in science and technology. The target is still so low, about 17 per cent of the age group, that it will probably not shake the essentially class character of higher education and with it the upper-class allegiance of all the administrative and technical cadres of the country.[8.30]

The concentration and militarization of science have effects that reach beyond the centres of research and production in the United States and Britain. The demand for raw materials by the United States is already draining the 'free' world; 22 per cent of the oil produced in Asia and South America now goes to the United States, which consumes 43 per cent of all world oil. In the same way the world is being drained of its best scientific talent. On the excuse, often genuinely believed, of assisting individual scientific work of promise, the best scientists, or at least the best of those untainted by communism or perverse patriotism, are being bought up and installed in admirably equipped laboratories in the United States, where they are free to pursue their own researches. This process, which began many years ago, is now reaching proportions that are endangering the scientific progress of many countries. Some half of the distinguished scientists in the United States are actually of foreign birth. Many of these, it is true, came to the country to escape Nazi persecution, but hardly any of them went back to their own countries after the defeat of Hitler. The gain to the United States in peace and war has been great, but it has to be balanced against a loss to the world. These scientists have been drawn away from the problems of their own countries at a time when their formative influence as well as their work is most needed. The drain of scientific research workers from other countries to the United States is now officially recognized. From Britain, for instance, a quarter of the physicist postgraduates go to the United States and few return. From India the proportion is not known, but nearly half of those that go stay. This trend increases the concentration of science in more highly developed industrial countries, lowering to a danger point the possibilities of the under-developed countries to push forward their own science and to catch up with them. The National Science Foundation estimated that between 1949 and 1961 44,430 foreign-born and foreign-trained scientists and engineers were admitted as immigrants to the United States.

The whole system of the concentration of science in laboratories which, however nominally under university control, are really under that of monopolies or government, and are directed towards projects deemed to be of military value, is a most serious danger to science.

Although the stringent conditions of security and loyalty now show signs of being relaxed, the general atmosphere is still such as to dissuade not only active but also prospective scientists from concerning themselves with the social implications of their work. Once they do so and American scientists as a body begin to assert their opinions with the strength to which their essential services to the country entitles them, we may expect to see great changes.

The relatively enormous wealth and productivity of the United States and the concentration of scientific effort there have correspondingly depressed the development of national scientific centres elsewhere in the 'free' world. Major research in nearly all fields, and particularly in physics, is now possible only in heavily endowed laboratories. These are now already largely to be found in the United States, and their creation elsewhere is an increasingly rare event. Already in the capitalist world only Britain, and to some extent Sweden, can claim to be fully independent in fundamental research, and that independence is precarious enough in several fields. Most other countries have governments in such chronic financial difficulties, largely due to military expenditure and trade restrictions, that they virtually starve science. For all the excellence of the work of their individual scientists, they are

381. In India old and new still form a great contrast. A gang of women labourers digging a canal to carry water to cool the country's first atomic power station. Photograph taken in 1967.

no longer able to carry out organized scientific work at a modern level, and tend more and more to be drawn into the United States orbit.

The growth of science in the capitalist world in the last few years has been phenomenal, but it has been at the expense of very serious distortions of aim and method. These have already alarmed thoughtful scientists, and by no means radical ones, on both sides of the Atlantic. Now there seems some hope, in an easier political atmosphere, that their voices may be heard.

SCIENCE IN THE DEVELOPING COUNTRIES

The criticism of present tendencies of concentration at the centre and neglect outside it applies with greatest force to the under-developed countries. Some of these, notably India, with a long tradition of science are nevertheless resisting them, and in the measure that they are achieving economic independence and building up heavy industry they are expanding scientific and technical education and research.

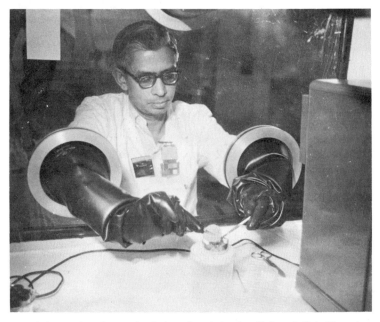

382. A large piece (100 gm) of plutonium being handled at India's Plutonium Plant. Note the radiation monitor on the operator's lapel.

There is still very little science in most of the remaining colonial territories, but a great popular demand for it. There the older imperialisms of Britain and France, to say nothing of Portugal, have been completely incapable of using science on a scale adequate even to exploit the natural resources. What science was used was devoted largely to the extraction of minerals and the production of cash crops, without regard to improving of the conditions of the peoples of the colonies. This is now becoming a thing of the past. The new states that are rapidly winning their political independence still have the greatest difficulty in developing their necessary scientific manpower to run the economy for the benefit of the people, and the advance that has been made has not yet been sufficient to do anything to narrow the gap between the developing countries and those of their industrialized former rulers. The only short-term solution is to build up the scientific and technical cadres in the under-developed countries with foreign help, at least until they are able to stand on their own. But so far only China has been able to do this. For help, to be really useful in building up these countries rather than exploiting them, must not carry with it economic or political domination. Help on these terms is now beginning to be available from the Soviet Union, notably in the installation of steel works, in the technique of oil prospecting and drilling and in the setting up of technical colleges. In sheer self-defence the capitalist countries may be forced to do the same and enter into competition as to who can help most.

THE LESSON OF THE ASWAN DAM

A telling example was that of the Aswan Dam. The United States refused at the last moment to finance the dam. Egypt accordingly seized the foreign-owned canal and the dam itself was in the end built with Soviet funds and technical assistance. Some years later, the great Volta Dam in Ghana was started with US support.

The events of the last few years should show, to all but those whose interests are not to see, that the colonial system in its old or new form is in any case doomed. Nothing can stop for much longer the upsurge of the peoples of the world to take over modern technology and science and to use the resulting wealth for their own benefit. The net result of this can only be an enormous gain of natural and human resources for the whole world. The scientific effort in particular will be multiplied.

Nor should this prospect frighten the people or the scientists of the old industrial countries. Their privileged position in a world of such grinding poverty is a curse and not an advantage to them. For the maintenance of that position is the reason, or at least the excuse given, for the crushing military burdens which bear specially hard on science. They are necessary, we are told, to check communism, the spread of which threatens the interests of civilization. In fact, once the artificial barriers to trade between the socialist and capitalist parts of the world are down, the rapid industrialization of backward countries would provide demand enough to keep the production goods industries of the old industrial countries in full production. And when, in a generation, the industries of the new countries had reached a comparable level, the standard of life of the peoples of those countries would be so much higher that they would provide an unlimited market for consumption goods. Helping backward countries is not a question of self-sacrifice on the part of the old imperialist countries, though they are, goodness knows, enough in debt already to the peoples they have exploited for centuries, but one of elementary self-interest.

SCIENCE IN THE SOCIALIST SECTOR

The contrast which the socialist countries offer to the rule of monopoly capitalism is an absolute one. General welfare and not maximum profit is the criterion for economic development, and consequently for the use of science. How this is done has already been described (pp. 1186 f.), but what has not yet been adequately stressed is the effect of this transformation on science itself and on its relation to the life of the people. The use of science for constructive plans in industry and agriculture demands an enormously increased output of trained scientists and a consequent increase in scientific education at the higher levels (pp. 1189 f.). Because of the practical interest in construction, in agriculture, in transforming Nature, in discovering and utilizing natural resources, in improving the health of the people, there is a much better balance of interest between the sciences, notably in the greater part given to geology, biology, and medicine.[8.48] One significant innovation is the entry of women into scientific work. In China and the People's Democracies, as well as in the Soviet Union, instead of the proportions of one woman to six men that is the rule in British science, as many and, in some fields such as medicine, more women than men are entering science. This means effectively doubling at one blow the intellectual pool from which scientists are taken.[8.14]

All this, taken in conjunction with the emphasis on science in primary school teaching, is leading to an enormous enhancement of the prestige and place of science in the popular mind. To see this one has only to compare the amount of space given to scientific affairs in the newspapers and periodicals of the Soviet Union and China with that in Britain and America.

What this trend is leading to is a radical transformation of the place of science in society, one which throws it open to the whole people, and not to the class-derived or class-selected *élite* who have monopolized it since the beginning of civilization. Such a transformation must bring enormous new strength to the countries where it occurs. In the effective competition which already exists between the two economic systems of the world, this throws into the balance new human resources, which, through science, can greatly speed up the utilization of natural resources. This has already gone so far that the Soviet Union is turning out more than twice as many trained scientific and technical personnel as the United States.

Once such competition in utilizing all, instead of a very small part, of human intellectual resources has really got under way, it cannot be stopped until the people of the whole world, and not of one class or country, are enabled by education and opportunity to contribute their full knowledge and ability to the general welfare.

THE ORGANIZATION OF SCIENCE IN THE SOCIALIST WORLD

In socialist countries, beginning with the Soviet Union, the organization of science has followed a different course from that in the capitalist world. Though military research is carried on, and successfully, as the achievement of the atom and hydrogen bombs show, it has neither absolutely nor relatively the prominence that it holds in capitalist countries.

Prominence is rather given to the use of science for the national economic effort. The need to secure the largest participation of science in industry and agriculture and at the same time develop science intrinsically, has been met, not by direct government action, but by an enormous extension of the older scientific bodies, and primarily of the Academies. The Soviet All Union Academy and its sister academies of Medicine, Agriculture, Architecture and Education, together with the newer Academies of the separate Republics, do in fact represent the ideal of the early seventeenth-century academies, like the Accademia dei Lincei, the Royal Society, and the Académie Royale des Sciences, brought up to the range and scale appropriate to the twentieth century. From being the nineteenth-century type of honorific society, the old Russian Academy found itself charged with the organization and running of large research institutes in every section of science, which by now have come to contain many thousand workers. Further, the

383. The building of the Academy of Sciences of the USSR in Leningrad. It was in 1907 that Ivan Pavlov (1849–1936) (see page 938) was elected President of the Academy and appointed director of its physiological laboratory.

Academy is responsible, through its own institutes and the directives it can give to scientific research in the universities, for the general planning of scientific work in conjunction with the plans of the economy as a whole.

The Academy's task is defined to be:

To determine the basic direction of research in the natural and human sciences and to direct and co-ordinate scientific research in these fields at its own institutes and those of republican academies and of higher educational establishments.

To promote research in pure science and in big general problems of application such as the use of mathematics and computer technology in economics, automation, new materials and new sources of energy.

To advise the government on the utilization of new scientific discoveries.

To direct the financing and equipping of the institutes of the Academy and the republican academies.

The Academy has been reorganized several times to meet the enormous changes of internal development and applications of science.

After the latest reorganization in 1963, its work is divided into three major sections which consolidate the twelve or more that existed before: 1. Mathematics and physics. 2. Chemistry and biology. 3. Social sciences. The sections will exercise general control over a total of some fifteen departments. Each department is to be a powerful and effective body, responsible for nation-wide development of its branch of science in the Academy itself, in the academies of the republics and higher education establishments. The inclusion of higher educational establishments is a somewhat belated recognition of the importance of the assistance which university departments can give to independent research institutes.

This system leaves the direction of science to scientists, the only people who are intrinsically competent to do it. It ensures, at the same time, that they have the means and the knowledge to develop science in the directions that seem to them richest in promise for the future. Contrary to what is often asserted, the scientific plan in the Soviet Union is not made for scientists but by scientists. Naturally they have the general economic plans in mind and know them intimately, because they have already been consulted in drawing them up. But these affect their own plans only in their broadest strategic aspects. The scientists will take, and they will be expected to take, far longer-term views. The great construction plans for the river valleys of the south-east of the Soviet Union, announced in 1950 to be completed in seven years, had been under scientific examination and discussion for twenty years before that. The annual and five-year plans of the Academy of Sciences have concentrated mainly on the growing points of science, though these are often also places where technical applications are most fruitful.[8·48; 8·69] A transformation is being effected in the socialist countries of the world of a kind that it is difficult for scientists in the capitalist world to understand. For to do so fully requires a knowledge and an interest not only in the science, but also in the history, the economy, and the philosophy of those countries.

The emphasis on an autonomous scientific direction, calling on the state for material support and answering its call for assistance in specific tasks, is the common pattern of scientific organization that has been adopted in the new People's Democracies and in the People's Republic of China. It has proved itself flexible and has released a wealth of ability and enthusiasm, recalling the great effort of national utilization of science that was inaugurated by the French Revolution (pp. 535 f.). It gives to the scientists greater powers, but also greater responsibilities.

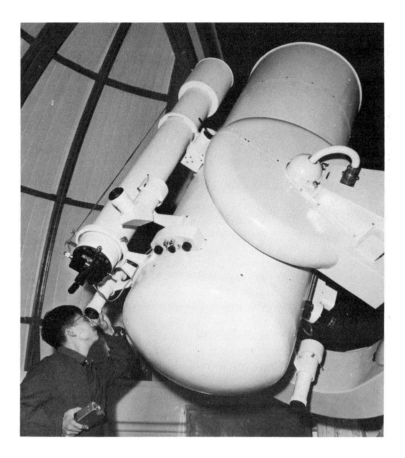

384. In the Chinese People's Republic scientific research is assisted by equipment constructed there. The manufacture of electron microscopes has already been illustrated (see illustration 236), but here a complex Schmidt-type telescope has been built in the workshops attached to the Tsuchinshan Observatory.

The scientific worker of the West finds it difficult to realize what is happening in Eastern Europe and in China today, not because the conditions as such are strange, but because they are the achievements of a people with a purpose: a purpose shared by all, including the scientific workers. When a common purpose exists the reaction of the individual is also transformed. The retreat to an ivory tower in science is in many cases simply an escape from the general meaninglessness and purposelessness of life in a world where the only prospect is destruction. Constructive social purpose carries with it emotional sanctions and satisfactions which we in this selfish civilization have lost, much to our detriment.[8.14a]

14.6 The Advancement of Science

This concludes the summary picture of the position and organization of science in the world today. It is against this background that we have to assess the discussions on general principles concerning the internal problems of science and of its place in society. These problems, which were posed at the very outset of this book, are in essence comprised in two questions. How can science be encouraged to flourish and grow? How can the results of science be used to the best purpose for the benefit of humanity? It was to find the answers to these questions, which are not merely academic but practical ones, that this whole inquiry into the place of science in society was undertaken. It can be justified only in so far as it helps to find them. The clues are contained in the actual story of science, if it can be read aright.

The way to answer the first question is to find the best conditions, external and internal, which have in the past helped the progress of science and to anticipate the changed needs of the present and future. The answer to the second question, which depends on the first, is set out towards the end of this chapter (pp. 1299 ff.). Some of the external conditions for the flourishing of science in the past have already been discussed (pp. 1220 ff.). In essence they are provided only in periods of social and economic advance, when science is given social importance and material means and is continually stimulated to new activity by problems presented to it from the economic and social spheres.

Now these problems have been essentially, as we have seen, those that touched the interests of the ruling class of the time, whether real, like navigation, or imaginary, like astrology. The opportunity and the honour given to the practitioners of science at any time are a measure of the degree to which they serve these interests. They are greatest in periods of active advance, because then the people who are occupied with science are closely in touch with the main economic interests, and are often drawn from the directing classes themselves or are brought into their counsels because of their abilities. We have had many examples in these pages such as: Archimedes, Grosseteste, Leonardo, Galileo, Boyle, Davy, Pasteur, and Kelvin.

But for the secure advance of science it is also essential that those interests should lie along lines which bring the scientist into connexion with constructive practical activities. For example, the relative sterility of natural science in the times between Pericles and Alexander (pp. 190 f.) is an indication of what able and gifted individuals could fail to do, if divorced from any relation to production through the overriding interest of the political problems of decaying, small-city, slave society. The Alexandrian revival of experimental science shows by contrast the immediate effect of contacts with the practical needs of a large-scale and, for the time, expanding economy.

THE SCALE OF SUPPORT FOR SCIENCE

The first condition for the material support of science is that it should be on an adequate scale. The progress of science has been halted over and over again by the lack of suitable materials, sometimes, it is true, because they were unobtainable, like rubber before the discovery of America; but more often because the scientists lacked the means to acquire them. Throughout most of its history science has been starved. The scientists have been driven to other pursuits to keep alive – like John Dalton having to teach children to read – and have found it hard enough to command the tools for their trade. Even now, in the era of government and industrial subvention of science, they are as often as not held up simply for lack of equipment. It is paradoxical when we consider the large part which purely scientific gear plays in our present civilization, where television sets and motor-cars abound, that the very people whose work made those things possible are often too poor to use them themselves, even for purposes which would result in their improvement. Scientific work is continually being hampered by lack of such elementary requirements as those of intercommunication and transport.

The total demands of science today, however big they loom compared to its penury in earlier days, are still small compared with the sums available for capital expenditure (p. 1253). Where the demands of science can be met, and where science is closely linked with an industry and agriculture which provide it with the necessary stimulus of problems, the external conditions for its rapid advance can easily be satisfied.

INTERNAL CONDITIONS FOR THE ADVANCE:
LANGUAGE AND COMMUNICATION

The satisfaction of the external needs of science should also make it possible for the scientists themselves to establish the necessary internal conditions for easy and rapid advance. The problem is to see to it that the individual scientific worker gets the conditions, opportunities, and incentives enabling him to give his best service. The work of science is social; it requires a sense of common purpose within each field of inquiry. It also needs mutual stimulations of its different fields, implying a good communication system and the absence of narrow specialization.

The technical side of these requirements is the easiest to meet because it is least complicated by external economic and political factors. Science needs to evolve a series of languages of its own, including mathematics, the general language of science. It needs logic and the capacity to formulate new ways of comprehending new things. Whatever the genesis of scientific ideas, they cannot spread or become fixed unless they can acquire a language suitable to them. That language may be geometrical or mathematical, that is, symbolic, or it may be the use of ordinary language in a special sense, that is, by the development of a scientific jargon. In both cases the purpose of the language is to establish a set of relationships which are understood in the same way by all competent people.

The difficulty is that the advance of science and its specialization make the relative number of competent people for any one particular scientific symbolism or jargon always smaller and smaller. There is a consequent danger that a scientific jargon may actually operate to hinder rather than to help scientific advance, especially when it is used to bolster up pretensions to special knowledge on the part of its adepts. In fact the progress of science has very largely been in simplifying and wiping out such specialized languages and replacing them by a common language (p. 38).

A RATIONAL INFORMATION SERVICE

Over and above the intrinsic difficulties of communication between scientists in different disciplines are those created by the multiplicities of language and the national barriers that today divide the world of science. These difficulties have grown enormously and have been made even worse by the growth of science itself. Already scientific papers of importance are being published in at least ten major languages, and over one hundred thousand almost entirely uncoordinated scientific journals are now being printed. There are as well endless difficulties about currency and security regulations. The situation has indeed been passed in many fields where it is easier to find out a new fact or build up a new theory than to ascertain whether these have been discovered or deduced before. It might seem as if the unity of science may break down under its own weight.

But this is in no wise necessary; however large an array of facts, however rapidly they accumulate, it is possible to keep them in order and to extract from time to time digests containing the most generally significant information while indicating how to find those items of specialized interest. To do so, however, requires the will and the means.

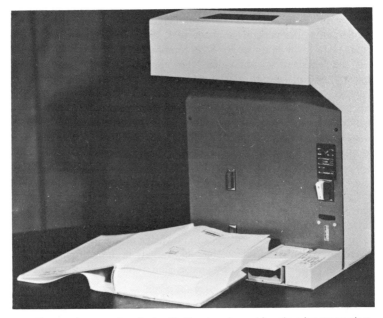

385. The growing quantity of scientific literature is reaching alarming proportions. Storage space is a problem and microfilm recording and reading equipment is coming into increasing use to save space.

Even twenty years ago it was evident that the communication system of science was entirely inefficient and wasteful and that it was only kept going by means of devices which actually made it worse, such as the founding of new journals, the introduction and circulation of reports and letters and, finally, the mere interchange of private letters. Thus the state of scientific communication in the most rapidly advancing fields of science has returned to what it was in the early Renaissance. It is this section that calls for the cause of most drastic action. The larger science gets and the faster it moves, the more important it is to know from one place to another what is happening and from one time to another what has been recorded. In the publication of papers we still follow the principle that science is a kind of Olympic or Nobel competition, the object of which is to reward the merit of outstanding people. This goes down to the lowest level in the thesis required for the first university post as an Assistant Lecturer. The result is a mass of papers commonly related by the private interests of different professors or those of business firms. To keep a record of these papers, there are abstracts which provide long catalogues, now almost impossible to buy and quite impossible to read, of all the papers. Next, there are a number of reports, some published by societies, some privately, planned to assemble and collate the advance of science in specialist fields. Finally, there are tables of data where the apparently valid results are codified, but these are usually years out of date and are not sufficiently critical. The result of this anarchic communication system, or lack of it, is that an enormous amount of knowledge is wasted and a great deal of duplication takes place. Worse than duplication – a number of promising ideas just fall out of count altogether. It is in the nature of the multifarious advance of science that ideas which appear in one field and may not have very much effect or relevance in that field might be extremely illuminating in another, but they hardly ever get there and the same discovery has to be made over and over again. It requires something more than the multiplicity of congresses, societies, and publications which organizations such as the International Council of Scientific Unions and UNESCO have tried to support and reduce to order. The realization must come, and the sooner the better, that scientists today must be prepared, to their own advantage, to spend some of their time in the service of arranging and disseminating information, and they must be enabled to do so by a financial support which might be up to 20 per cent of the cost of the research. The idea that an information service, outside the technical field, where money is no object, can pay for itself is no longer valid. Only governments can

386. The greatest effort at co-operative scientific research after the Second World War has been the International Geophysical Year. This ran from July 1957 to December 1958 and scientists from more than sixty nations took part. This stamp was issued in the United States and shows, over hot gaseous prominences from the Sun (the flame-like shapes), a pair of hands stretching out to each other. The Sun was the body that caused most of the geophysical effects being studied, while the hands are a reproduction of the hands similarly stretched out in 'The Creation of Man' by Michelangelo.

afford to run comprehensive scientific information services, but these will prove to be the most economical because they avoid duplication in such common features as intake of journals, mechanical handling, automatic retrieval and translation (pp. 855 f.). This has certainly been the experience of what is probably the largest such system in the world, that of the Academy of Sciences of the Soviet Union. At the present time, when international scientific relations are being resumed in a better atmosphere than at any time since the war, it would be appropriate to conclude what was started by the Royal Society in the Scientific Information Conference of 1948 – a serious attempt to provide world science with a comprehensive and up-to-date information service.

THE UTILIZATION OF INTELLIGENCE

However well the technical problems of communication in science are solved, the basic internal condition for the flourishing of science remains the human one. Science is ultimately the work of many individuals of different grades of ability, for the achievements of the greatest minds in science could never have been reached had it not been for the patient and accurate work of hundreds of others who had no need of great imagination or synthetic grasp.

Now there can be no real shortage of intellectual ability in the human race. What has been achieved already in science and culture has been the work of a small handful of men, drawn from one small social class of a very few cities and nations. By calling on men and women of all classes and peoples we should be able to multiply manifold the rate of advance of science. To do this, however, implies an altogether new attitude to popular education, in which, following the example of the Soviet Union, a secondary and soon a higher technical or university education will be made available to all. That education must largely be based on science, in the broadest sense, social as well as natural. It will provide at the same time the field from which will come the research workers who will build the science of the future and the workers in other fields who will have learned enough of the significance of science to co-operate actively with the professional scientists (pp. 1189 f.).

Science will not fail for lack of human capacity; where it fails it will be for lack of social organization to make use of that capacity. The brilliant, the capable, and the industrious must all be enabled to come into science, and in such a way that they can all give of their best in its service. They must be imbued with a purpose and a conscious feeling of working for an end of which they approve. The inspiration of the scientists of the great period of advance was the belief that they were working for the common benefit of society. That inspiration is in danger of being lost in a culture the ends of which are private profit and war, now that these are seen to have lost any of the justifications they once may have had.

CO-OPERATION IN SCIENCE

Individuals, however inspired, do not do their best work alone. The peaks of advance in science have occurred when a number of people have already been actively working together in each field of science. This, through mutual suggestion and emulation, increases enormously the chance of hitting on significant discoveries. Equally important has been the effect of one science upon another. This has occurred largely by the importation into one field of knowledge of ideas derived from another, sometimes directly, sometimes by analogy. The proved value of the analogical approach to scientific theories stresses still further

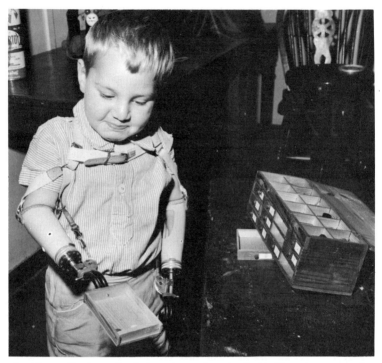

387. The applications of techniques in one branch of scientific endeavour can be of great assistance to another. Biological engineering is a case in point and a child born without hands can now be fitted with artificial equipment with which he can become quite dexterous.

the unity and interrelation of the different sciences, and exposes the sterility that accompanies specialization. Most fundamental advances in particular sciences have in fact come from individuals who have had relatively little experience in that science. The revolution in chemistry in the eighteenth century was effected by Priestley, who was no chemist (p. 531) and it is notable how great advances in medicine have mostly come from men like Pasteur (p. 647) who were outside the profession, while the other sciences have gained enormously from the spare-time work of doctors like Joseph Black (p. 580). The analogies creating a new scientific theory have usually come from science of a simpler character than those they are applied to, as, for instance, Dalton's atomic theory, which came directly from a consideration of Newton's particle dynamics.

The failure of communication, already referred to, tends to reduce the work of discovery to an individual pleasure, quite independent of whether what is discovered is new or even whether it is true. However, a complete scientific life is impossible without a justified consciousness of the value of the work, if not to the community, at least to the world of science. That justification can only be found if that world really works as a co-operative unity. It must be an open one not crossed, as at present, with barriers of 'security' and accessible only to 'screened' personnel. It must contain no 'iron curtains' or 'classified' fields of research.

The ill effects of secrecy are of two kinds, the open and the concealed. To work in the knowledge that relevant information is not available is absolutely damaging and subjectively discouraging. Not to be able to divulge what one is doing so as to get the advantage of others' experience and criticism tends to narrow and slow down the work even more. Both these tendencies are exaggerated if, as happens especially in military research, security is added to secrecy and the whole of the worker's life, character, and associations come under continuous and secret observation. Not only does this drive people of independence and ability out of such service, but it also psychologically warps those that stay in, even when they are inspired by a genuine belief in the rightness of the end for which the research is carried out. This process has been exposed to public gaze in the case of the citing of Robert Oppenheimer as a security risk.[8.19]

The concealed effects of secrecy are necessarily impossible to assess. What is lost by separating elements of knowledge can only be guessed. But it is certainly a large loss, as is shown by the rapid advances that are made when two different aspects of science have come together, as in the discovery of electro-magnetism. Other evils of a more human nature also supervene. Work which may not be published cannot be effectively criticized. Incompetence, jobbery, private intrigue flourish under the cover of secrecy. The opportunity of denouncing one's rival as a security risk always exists. Every tendency inimical to the pursuit of science is encouraged. As a French general once remarked to me: 'Le secret militaire c'est fait non pour cacher les connaissances mais l'ignorance.'

14.7 Contemplation and Action

THE PLACE OF PHILOSOPHY

So far in this discussion we have been concerned with science treated as if it were some kind of autonomous and independent entity, though one capable of being affected by outside circumstances. It should be clear from earlier chapters that this is only a very partial and, by itself, misleading picture. It is only for the purposes of argument that science can be considered separately from the society of which it is a part. The influences from society have not in the past, and do not in the present, react on science only from the outside, but affect its internal constitution and activity deeply and directly. Questions like the place of philosophy in science, the balance between freedom and organization, and the moral responsibility of the scientist, all relate to internal difficulties and conflicts of science brought about by the conscious or unconscious operation of social forces.

In their first appearance science and philosophy, as we have seen (pp. 171 f.), were not distinguished. The Greeks, who forged the term for both, considered that they both served the same purpose. They contained the abstract knowledge of the history, construction, and operation of the universe, knowledge to be achieved by natural or supernatural means and to be treasured for its own sake. This is essentially a magical attitude towards science, one that persists in our own time. It provides a very convenient cover for those who profit from science, enabling them to rule out of court as low and materialistic the idea that it should be used for human welfare.

The early attitude towards knowledge was contemplative rather than active, consonant, as has been shown, with the monopoly of knowledge by *élites* free from the cares and experience of manual work, first by administrators, then by privileged citizens, latterly by churchmen. Because of their interest in the *status quo* of society they preferred to think of knowledge as a static perfection, obtainable by reasoning from simple observation or resting on sacred revelation. To attempt to change it was not only vain and unnecessary but downright impious.

This attitude could not be maintained in the face of the economic and technical changes that accompanied the development of civilization and its subsequent transformations. More things had to be taken into knowledge, and knowledge had to be more effective in connecting together the things that were known. The active mood superseded the contemplative one. Ever since the Renaissance it has been recognized that science is not static, that the gaining of new knowledge rather than the affirming of old is the essence of science. Even up to the present however, it is still tacitly assumed that this is in some way an exceptional process, that the objective is the discovery of some final truth about the universe, the contemplation of which is the end of science.

This is the very attitude that has preserved as long as possible all the old and now meaningless forms of philosophy and theology (pp. 1160 f.). Its influence on science is equally dangerous, but is much more effectively concealed because in the presentation of science the under-lying philosophy is taken for granted and is nowhere explicitly stated or criticized; it finds little or no place in the literature of science. A scientific publication is considered adequate if the observations and experiments, the conclusions and the arguments leading from the first to the last, are clearly stated. True, this is all that is necessary for the immediate transmission of scientific knowledge, enough to enable the work to be reproduced and for variations to be made on it. But science has longer views than this. What is not stated, and what may be of greater importance for the future of science, are the reasons why the work was undertaken in the first place, and the record of the actual, as against the rationalized, train of thought that led to the inferences that are made. These are omitted, because the former is considered irrelevant and the latter too difficult or perhaps too trivial to put down. This does not, of course, mean that philosophical ideas are not passed on in scientific literature. What it means is that they are passed on in an unconscious and traditional way, so as to perpetuate inside science attitudes and prejudices from the past, always heavily biased in favour of ruling-class interests.

The omission of explicit philosophy from science is not an accident; it had a good historical justification, though one that no longer holds. The philosophy of the Ancients and the schoolmen was adapted to religion and politics, and not to the material handling of Nature. It was a hindrance and not a help to science. But it was impossible for the early scientists to attack that philosophy directly; they had enough to do to be allowed to carry on with their experiments in safety. It was better to ignore it. Moreover, the greatest upsurge of science occurred very largely in Britain and Holland at the time of acute religious and political difficulty, when it was a matter of elementary good sense not to discuss philosophic matters. Consequently the tradition has grown up, and is so ingrained in British science and, through British science, in the science of many other parts of the world, especially that of America, that philosophy as such has no part in scientific matters: that, in Newton's motto, *Nullius in verba*. Science is deemed to proceed by way of man's common sense and practical understanding (p. 441).

What we are now beginning to recognize is, that while it is intrinsically impossible to maintain and develop the discipline of science without an underlying tradition, this evasion ensures that the tradition should be tacit and unexamined. All that the neglect of philosophy will do is to hide a great deal of very bad, outworn, and untenable philosophy. A further consequence is that the lack of capacity, means, or time for thinking about the fundamentals of science will hold up its progress and keep it along accepted channels until conditions become so unsatisfactory that it has to break out into new paths by the accident of discovery, instead of being able to arrive at them by any rational process.

It may appear in these remarks that I am ignoring the multitude of works that have appeared in the last 300 years on the philosophy and method of science, from Locke and Hume to John Stuart Mill, Pearson, and Eddington. These are certainly contributions to philosophy, but they are concerned with only extremely limited parts of science, mainly mathematical physics, and are in no sense a philosophy of science as a living whole. Very few productive scientists read them, they are hardly ever quoted, and it is difficult to find a single example where they have led to a discovery or an explanation of any scientific fact.

The reverse has often happened; as we have seen (pp. 433, 476), both the explicit and the implicit philosophies of science have in the past acted as limiting rather than as liberating factors to scientific advances. The greatest advances of science have been made in spite of them and not because of them. The more the advance of science is kept clear of such unnecessary obstacles, the greater is the opportunity for those of ability to pit themselves against the real and not the artificial difficulties, and thus to make sweeping and planned advances in science where previously it has moved forward with halting steps. This does not mean that philosophy must be discarded in science, quite the contrary. In criticizing current philosophies it becomes evident that they fail through their partial, unsocial, and unhistorical approach to the underlying problems of science and also through their ruling-class bias, largely unconscious because so completely taken for granted.

As already indicated (pp. 570 f.), the form this interference takes is a tendency towards positivist, idealistic, and formal philosophies of science, which in effect draw the scientist away from an active experimental approach to problems into a passive and contemplative one, either immersed in a stream of meaningless or unreal experience, or reflecting on eternal and abstract truths. These paths in their ancient and modern forms only lead to sterility, and science has escaped from them in the past only through the effect of impacts from the material and social world, which have shaken its complacency and forced it to face new problems. Any valid philosophy of science must take these facts into account; it must see science no longer as static and isolated perfection but as part of a changing, real, material, and social world. The first steps towards such a philosophy were charted by Marx and Engels many years ago. Subsequent experience has deepened and broadened their conclusions (pp. 1167 f.). This is not to say that such a philosophy for the natural sciences has yet been hammered out. That is a task for the future. It needs to be derived not from any abstract and *a priori* logical analysis, though logic must be an essential item, but from an active experience of using science in relation to its social tasks.

PHILOSOPHY IN THE NEW SCIENTIFIC REVOLUTION

It is certain that the enormous changes of the last decade, which are already clearly and officially recognized as the new scientific and technological revolution, are also bound to have a decisive effect on the development of philosophy. The great advances of molecular biology, with the renewed interest in the origin of life, and the enormous capacities for substituting or even transcending human thought, which electronic computers have brought, must find their place in the new philosophy. We may even categorically state that no philosophy that does not take these advances into account may be seriously considered as having any other than historical interest.

The change is bound to go deeper than the inclusion of new pieces of information and the new outlooks they bring with them. The realization of the *process* of science itself and especially of its extremely rapid advance is bound to put into question the very central objective of philosophy – truth. First in the course of a single lifetime, then in a decade, now in a year we have found new basal facts and whole attitudes of viewing the universe are going to change. There is no evidence at all that this process is slowing down, on the contrary, it is speeding up. This fact itself must be recognized and is in effect being recognized in practice by most of the scientific thinkers who have added to our knowledge. I have called it *provisionalism*, others will find some word less tough and long-winded. It is more than scepticism, it is a conviction that whatever we think now, people in a very short time from now will think differently and better. Truth thus presents a moving function:

we happen to believe, for the time being, fully recognizing that it *is* only for the time being. The search for more knowledge goes on, but that for a stable complete knowledge must be definitely abandoned.

14.8 Organization and Freedom of Science

The great developments in the scale and organization of science in recent times have had a direct effect on the internal character of science itself. Scientists have been obliged for the first time to take stock of their activities from a general sociological and not only a particular academic aspect. They have to consider their relations to each other and to society, as well as to the subject of their own researches. And this interest is not confined to the scientific workers; it is also a matter of the highest public importance. The more it is realized that the actual welfare and future progress of society depend on the appropriate development and use of science, the more people will be willing to support and foster it, but at the same time the more they will be concerned to see that science is healthy and effective.

Ultimately, however, it is only those inside science who can find out how to do this in detail and what degree of outer support and co-operation is needed. Not unnaturally, in such a period of transition there are wide differences of opinion. Two major contentions are at present agitating the world of science: Is the organization of science compatible with that freedom under which alone science can advance? Are scientists responsible, and in what degree, for the social effects of their work? These are effectively two aspects of one problem, and the resulting arguments have divided scientists into two fairly definite opposing camps. The older school, looking back to the golden age of nineteenth-century science, wish to reduce organization to the minimum to allow for the free and spontaneous effort of individual and personally devoted scientists.[8.6-7; 8.53] They also wish to dissociate themselves as far as possible from any responsibility for the effects of science, which they are willing to pass on to the industrialists and politicians, though for the most part they affect to deplore the consequences. On the other side are to be found mostly younger scientific workers, who see in organization the only means of advancing science and of securing its effective use for social ends. These scientists feel that they must take their share of the responsibility of the use of science in society as part of a wider democratic movement.

It is here particularly that the contrast between the use of science in capitalist and socialist countries has had the greatest effect on the opinion of scientists. On one side, all can see the great development of industrial science for the profit of monopolies, and the even greater development for military purposes, with science playing a predominant part in devising weapons of mass destruction. On the other side there is the creation of a new large-scale organization of science which, though military science has its place, is directed primarily at solving problems of production, and at setting on foot new constructive schemes for the changing of Nature and the raising of the standards of living.

Already during the Cold War period, despite the virtual suppression of information from socialist countries and a violent propaganda campaign magnifying every absurdity and failure, scientists in capitalist countries were beginning to doubt the picture they had been given. They found it difficult to square with the undeniable fact that Russia

388. Some organization of science is already a *fait accompli*, especially where the subject of research is, like nuclear fission, full of potential dangers to the public in case of accident. The experimental nuclear reactor at Dounreay using a liquid sodium–potassium alloy as a coolant is a case in point. Here research is to investigate the possibilities of a fast reactor and the technology arising from it.

had, from a very poor beginning, become in thirty years the second industrial country in the world and that Communist China in five years was well started on the same path. These intimations became certainties after the Geneva Conference on Atomic Energy in 1955, where scientists could compare notes for the first time. Since then reciprocal visits are rapidly dispelling prejudices as to the incompatibility of socialism and science. Even if they do not like everything they see or hear about, they admit that there is something serious there and that it is growing very fast.

Such comparisons reinforce the scientists' concern, especially in Britain, for their own conditions of life and work and for the use to which the results are put. They see promising research programmes – except when they are of military interest – being indefinitely deferred. They note the inadequate and dilatory application of science to industry and the blocking of scientific education by inadequate buildings or salaries for teachers.

These disabilities are felt by more than those immediately affected by them. An ever larger public is becoming aware of the rapidly growing possibilities opened up by recent scientific discoveries. They can see for themselves that they are not being followed up, and that the present halting advance of science is far less than it might be. They sense that they are being cheated of their birthright, the fruits of knowledge, of which Bacon spoke at the dawn of the capitalist epoch. They begin to feel that if this system cannot use the gifts by which it grew, it may be time it was reformed or gave way to a better one.

INTRINSIC PROBLEMS OF THE ORGANIZATION
OF SCIENCE: ORDER AND SPONTANEITY

Even for those scientists who have not been inclined to push their analysis of the position of science as far as this, or who may accept it as satisfactory, the problem of the best way to organize science cannot be entirely avoided, for in their day-to-day work they needs must find themselves dealing with organization such as it exists today. They may disapprove of it in principle, but in fact they cannot do without it. That disapproval has, indeed, some rational basis, in so far as the problem of the organization of science differs radically from that of organizing nearly all other human institutions from war to commerce and even sport. Only the organization of art offers greater difficulties.

The reason lies in the fact that science differs, as we have seen, from other human disciplines, in that it deals with the new and not with the expected. In other fields it is possible to determine what can be done and what steps are necessary to do it. Not so in science. In dealing with the unexpected, something quite different from routine competence is required. There is, of course, plenty of routine in science, and of a proportion necessarily growing with the scale and intricacy of scientific techniques. This routine is essential, and science could not exist today without services of technology, supply, administration, and communication unthinkable in earlier ages. But no scientist imagines that this necessary complement is a substitute for the central characteristic of science – original discovery.

The crux of the problem is how to secure the conditions necessary both for the material continuation of science and for its capacity to find new things. A splitting of science into two parts, one applied and routine, the other pure and free, is no solution. Indeed as they are two aspects of the same organism, like seeing and moving, it cannot be done. The history of science shows that at all stages new aspects of Nature have been discovered in the process of solving practical problems, and, conversely, that practice withers and decays if not revivified by abstract thought (pp. 585, 623).

THE FLIGHT INTO ANARCHY

It is even more impossible to restore anarchic freedom to the whole than to a part of science. It is, of course, understandable that in the reaction to the often stupid and autocratic direction of science in industry or in war there should be a movement to escape from any organization whatsoever. But the scientific flight to the desert for solitude and meditation is an intrinsic absurdity, for science is of all human enterprises the most dependent on mutual assistance and understanding. Science has never been really free, but whatever apparent freedom it has had belongs definitely to the era of competitive capitalism, and it is completely incompatible with the new industrial revolution, and its large-scale, organized production. It is, in fact, as much an anachronism and a freak as the mock Gothic of millionaire-sponsored universities. The idea behind this striving to escape is one already much discussed in these pages, that of the scientist as a dedicated person above the common struggle. Because the scientist is traditionally the successor of the man of learning of the past, who felt himself to be socially, materially, and intelligently above the mere technical hand worker, there has arisen the conception of the scientist as one of an *élite*, a set of people who are apart from society and supported by it in order that they should, through the exercise of pure thought, bring a reflected glory of achievement to the populace that cannot reach these high intellectual levels.

From the earlier stages of civilization until today this conception of a learned *élite* has been one of the strongest supports for the acceptance and perpetuation of a class society. Where it has been most marked in the past it has led to the stagnation of science, cutting it off from the stimulus and check of practical life and turning it into useless and repetitive pedantry. Nevertheless this idea has many supporters today, particularly among scientists of age and eminence.[8.2; 8.20; 8.22] They see in the present political conditions of western Europe and America the only means of preserving a position for scientists without which they cannot conceive of science existing.

These defenders of the freedom of science, though often unconsciously, are even more concerned with the defence of capitalism as a way of life against the new socialist way which is challenging it. They find it difficult to accept the idea of working for any organization, even when it is deliberately directed to the common good. They resent most vehemently the new responsibilities and the new ideas that they believe are being imposed by force on scientists in communist countries.[8.42] They much prefer the freedom and irresponsibility of an unorganized system where individuals can seek for knowledge and happiness in their own way. The fact that the need for organization has arisen out of the failure of anarchy is hidden from them, because of their profoundly unhistorical attitude towards society and their own work. It is significant, however, that their resentment of organization does not apply to the organization of science in the pursuit of the private profit of the large monopoly firms which between them control nine-tenths of industrial science (p. 1254), nor to the action of the governments which utilize nearly all their scientific resources for war preparations (pp. 846 f.). These abuses they find relatively easy to tolerate, provided that some islands of non-interference in which a few scientists can pursue their private researches are preserved for them.

THE SEARCH FOR A SOLUTION:
THE INNER DEMOCRACY OF SCIENCE

However impracticable or reactionary these solutions are, the problem that they purport to solve is real and important. What we need to evolve is an organization which makes use of the greatest advantages that can be drawn from co-operative action, while preserving the advantages that belonged to the older period of unorganized science.

It must combine the character, flexibility, and individual initiative of the science of the past, with the team-work and strategy needed to cope with the vaster problems of the future. Whether this problem can be solved at all in the framework of capitalism is an open question. Indeed it seems most unlikely that capitalism will ever be able to adapt itself to the requirements of a full utilization of science. The steps that are taken in this direction lead to nothing because of the overriding drive for profits and to war. But that is not to say that it is not worth while trying to solve it. Any gains, even partial, in the way of a freer and better organized science – for these are complementary and not mutually exclusive terms – is a social gain.

Freedom can be combined with organization by the greatest measure of informal co-operation and inner democracy. This is in full harmony with the traditions of scientific work in capitalist and socialist countries alike.[8.11] It fits especially well with the tendencies already apparent in those parts of modern science where the greatest progress is being made, notably in physics and biochemistry. Teams are replacing individuals, and the problems of co-operation are being faced in actual day-to-day practice in many modern laboratories. There are undeniable difficulties. The temperamental attitude of many scientists is averse to co-operation. They have in fact been selected, or have selected themselves, because of interests which have to a certain extent separated them from their fellows, while the desires of being alone and being given sole credit still often interfere with and bedevil attempts at scientific team-work. These attitudes, however, have in the past been powerfully reinforced by the social atmosphere of competition and personal advancement; they are far less inborn than people used to think, and the experience of the satisfaction of team-work will go far to dispel them.

To talk of the inner democracy of science is not to imply that it is an isolated democracy: that scientists should run the show regardless of the world outside. Quite the contrary. The democracy of science is conceivable only as part of a wider democracy. Between the two a large degree of interpenetration would be needed. It is not sufficient, as the experience of the war has shown, for administrators or service chiefs to pose problems and for the scientists to solve them. The scientists have to be among the administrators and technicians to find the problems in their real contexts.[8.11] Conversely, scientific research itself needs its administrators and technicians. Nor should any organization of science give further support to the concept of scientists as a separate *élite*, even if working for an accepted common purpose. Scientists are just one kind

389. Education in science is now becoming recognized as a vital factor in the education of all. With this recognition has come the appreciation that modern technology can add materially to teaching methods. In particular, electronic equipment can be used – in this case to help in the learning of a language in a language laboratory.

of worker – they are essential but not superior to any other. The new realization of this vital function of the scientist has now reached official levels, as witness the concern with scientific organization and teaching. We must consider the present period as one of transition, and dangerous on that account. As long as the direction of society is in the hands of those educated for a previous world, and consequently incapable of knowing what they are doing or ordering at a time when the technique of destruction has been built up with fatal completeness, no one can feel safe. This is not a plea for scientists in control, but rather a plea for the general education of an increasing number of people who will be scientists and the abolition of the distinction between physical science and other forms of culture which has grown up in this century and needs to be abolished.[8.59]

THE STRATEGY OF SCIENTIFIC RESEARCH

The previous considerations apply mainly to the detailed running of science, its *tactics*; but organization limited to this would not have much effect.

Science needs a *strategy* as much as tactics, in fundamental as well as in applied science. The tendency to specialization in the sciences – the development of jargons, the much longer training needed to understand branches of science – make it imperative to provide some means of linking them so that they can be of use to each other, rather than each following its separate trend. In Britain, at least, it was the urgent needs of the war that gave its fullest impetus to the concept of a strategy for science. There science had to be directed for definite outside purposes, each of which involved nearly all the sciences. This tended to break down the barriers between them, and led to the idea of 'objective' or *convergent* research, in which the resources of all disciplines were directed towards a common 'object': the solving of a technical or operational war problem such as the piercing of tank armour or the defence against submarines. Complementary to this was the development of 'subjective' or *divergent* research, where use was sought in all fields for the product of one particular set of devices or ideas, such as radar or planned maintenance of aircraft.

The value of these methods of procedure in war, under the general title of operational research, appeared so great that it was natural that those with some experience of them would want to use them in peace, where, it is true, the objectives were much vaguer but were at least constructive. How and why that attempt has led to nothing has been discussed in Chapter 13. This failure, however, was entirely due to the conditions imposed by a capitalist economy; the need for a strategy of science still remains, and its possibility in practice has been amply demonstrated in socialist countries. The attempt to use strategy in science implies a new dimension to scientific thought; a need to consider the whole progress of science, not merely one's own part of it, and to relate science not only to contemporary observations but to its past and its future.

THE ORGANIZATION OF SCIENTIFIC WORKERS

Transformations in science cannot come about by themselves, even under the stress of external objective conditions. Like those in every other sound institution, they have to be effected by actual men and women inside and outside science, who from their knowledge and experience of the trends of events have seen the need for intelligent, co-operative action and who are prepared to devote themselves to it. In the first place this drive comes from inside science itself. The enormous growth of science in the nineteenth and even more in the twentieth century has brought into being a new and rapidly growing profession. This new profession of science is radically different from the profession of science in the nineteenth or earlier centuries. There are probably now in the world some 500,000 men and women who actually make their living by scientific work, and among them some 100,000 scientists who are occupied mainly in research. The vast majority of these are salaried employees of industry and government, and only a small but vitally important proportion are found in academic bodies. The growth, however, of the number of scientific workers has been so rapid that at first it outstripped their possibilities of organization. The older scientific societies, concerned primarily with the internal advance of science and secondarily with the setting up of professional standards, were effectively incapable of providing such organization, nor in fact did they wish to do so.

The new type of organization that arose first in Britain and is now spreading to other countries in the world is one of a frankly trade union character, admitting the existence of science as a new factor in industry and agriculture, and of the scientific worker as another but not essentially different kind of technical worker. In Britain this resulted in 1917

390. In recent times there has been a tendency for scientists to be organized in trade unions. In Britain in 1917 the Association of Scientific Workers was formed, and in 1947 an international association, the World Federation of Scientific Workers, was founded. This photograph shows the Second Meeting of the Executive Council of the World Federation, held in Paris in November, 1946. Sitting at the head of the table (third from left) is Professor F. Joliot-Curie, the Federation's Founder President, and next to him (fourth from left) is the author, Professor J. D. Bernal.

in the formation of a special trade union, the *Association of Scientific Workers*. In other countries with a different trade union structure, in which all workers in a particular industry belong to one union, it has resulted in an assembly of scientific workers from different trade unions who consult on their common problems. In countries with weakly developed trade unions, or where scientific workers would not be likely to belong to trade unions, independent associations of scientists have been formed, whose objects are generally limited to securing an adequate place for science in their countries. Most of these organizations are now associated together in the *World Federation of Scientific Workers* founded in 1946.

The objects of such associations are twofold: first, as trade unions, to look after the interests and conditions of work of their members, a most vitally necessary task as the unorganized worker of today is most ineffectively protected against exploitation; and, second, to concern themselves with the proper utilization of science in the national economy and international affairs. As the preamble to the *Charter for Scientific Workers* puts it:

> Scientific workers can adequately carry out their responsibilities to the community if, and only if, they are working under conditions which enable them to make full use of their gifts.
>
> The primary responsibility for the maintenance and development of science must lie with the scientific workers themselves, because they alone can understand the nature of the work and the direction in which advance is needed. The responsibility for the use of science, however, must be a joint responsibility of the scientific workers and of the people at large. Scientific workers neither have nor claim to have the control over the administrative, economic and technical powers of the communities in which they live. Nevertheless they have a special responsibility for pointing out where the neglect or abuse of scientific knowledge will lead to results detrimental to the community. At the same time, the community itself must be able and willing to appreciate and to use the possibilities offered by science, which can be achieved only through the widespread teaching of the methods and results of the natural and social sciences.[8.71]

Other organizations of more 'establishment' and less trade union character, but still concerned mainly with the social effects of science, have grown up in the post-war period. The best known of these are the Pugwash movement (pp. 846, 1163 ff.) and that of Linus Pauling, which concern themselves with the scientists' responsibility to reveal the dangers of modern war and to urge the use of science for constructive purposes.

THE POPULARIZATION OF SCIENCE

This implies a community of thought between scientists and non-scientists, a wider understanding of social problems on the part of the scientists and a wider understanding of science on the part of administrators, workers, and people at large. The process of spreading the knowledge of science is one which needs to be resumed with a far greater intensity than it had even at its heyday 100 years ago. But the new popularization of science must differ from the old. It is no longer a question of merely exhibiting the wonders and potentialities of science. 'Let us have more science,' said the Huxleys and Tyndalls, 'and then life will be safer, wealthier, and more agreeable to all.' That is now not enough. In fact it is necessary to deal with the very real suspicion and even hostility to science that has been generated in recent years by the use of science for destruction, and the universal fear of the further use of science in more and more horrible forms of destruction in the future. A widespread understanding of the relation of science to social progress and a determination to act on it will be needed before science can be made safe for the world. For the full, positive use of science far more than a passive acquaintance with science is required. Once science forms part of general education, active participation in science by all the working population (pp. 1189 f.) becomes possible and indeed necessary. Every phase of the productive process in industry and agriculture, even the practical aspects of domestic life, can become a field of intelligent experimentation, and practical improvement and innovation.

14.9 Science in a Rapidly Changing World

The character of the twentieth-century scientific revolution poses new problems for scientific organization, some of which have been already treated. In many respects they differ in kind from any that science has had to deal with before. The problems are certainly greater, but in face of that we are very much more conscious of them and ought to be able to plan how to deal with them.

The new conditions primarily include the effects of the very size and cost of science, of its extremely rapid growth and, in consequence, the need for the education of scientists, in fact, for the education of the whole population in science. Then we have the emerging needs for the planning of science for the creation of a science of science.

We are only at the beginning of the period in which science is the determining factor in economic and cultural life, but we are still hampered by the traditions which grew in a period when science had very little place in life and where it was treated as an optional extra. The very size of science today and the number of people engaged in it at every level, from the advancing edge of science in research to its application in an increasingly widespread practice, is only a foretaste of an age of science to come, when it will become the central factor in production, agricultural as well as industrial. Already, especially in technology, the scientific mode is steadily replacing traditional modes. This size factor is posing problems of its own and although science advances it is definitely becoming much less efficient. It is time to stop thinking of science as a cultural factor in civilization and to think of it

391. Since science is now the determining factor in economic and cultural life there is a pressing need for the established scientist to make known the results of his research, and to review his work in a wider setting. The medium of television is becoming increasingly used in this respect. The photograph shows Professor Herman Bondi during one of the five transmissions he made on the subject of gravitation.

more as an industry, applying to it criteria of efficiency and demanding adequate organization.

Another consequence of the size of science is the way in which it encompasses not only all kinds of activities but also spreads its activities all over the world. Increasingly, science is becoming world science, occupied with problems such as meteorology and oceanography, which have no meaning except on a world basis, and also with problems of geology, mining, petroleum and water resources, which are also world problems. Further, actual scientific discovery and its application is no longer concentrated in a corner of western Europe and north east America but is rapidly spreading. We have to envisage a world in which science will be uniformly distributed and produced to about the same extent in all parts of the world.[8.55; 8.58]

THE NEEDS OF SCIENTIFIC ADVANCE

Not only is the speed of scientific advance itself very much greater than it has ever been, but its rate of growth is also increasing. This results in an almost paradoxical situation in which it is practically impossible to keep up to date and, therefore, to produce effective work in science in anything but a small field. This is a paradox, because the nature of modern science is one in which interrelation between different fields is greater than it has ever been and where it is therefore more important than ever to have information of all different kinds of science at the same time. The consequent evils of ignorance and duplication of effort are also growing. But if this is the case at the surface, it is something that also goes very deep. New discoveries are not assimilated – there is no time to do so – and the whole ideological structure of science, scientific theory and hypotheses is in a state of permanent and growing confusion. The result is that we can no longer talk of a *state* of science because it is moving too fast and the approach that I have discussed elsewhere (p. 1281) as provisionalism becomes a necessity. This further implies the need to take deliberate steps to keep science as well-ordered as it can be at all stages, to point out where the difficulties are and to see that people deal with them. The attachment of theoretical scientists, not only in regions where theory is indispensable such as high energy particles or genetic codes, but also in all branches is becoming a further necessity.

SPEED OF GROWTH AND EDUCATION

One aspect of the speed of the growth of science is its relation to education. It is quite evident now that there is little hope for anyone who

392. The television lecture has also been used in the field of molecular biology. Professor J. C. Kendrew is here demonstrating a model of the protein myoglobin. For his determination of the structure of myoglobin Professor Kendrew was awarded a Nobel Prize in 1962.

has ceased learning it in the university for even as much as ten years to keep up with the advance of science over the wide field which has now become essential to an effective understanding and practice of science. We have to recast our education in depth. Clearly, more education and longer education is needed. This will fit in with the release of human manpower due to automation, even in the clerical professions. But it introduces a new concept, that of continuing education throughout the whole of life. It is a matter of adapting the Chinese model of 'learn while you work and work while you learn'. This is in harmony with the general breakdown of the old, definite age groups based on manual work and child-bearing and rearing – 'the seven ages of man'. The educational system will have to turn into one in which much greater importance will be attached to the education of men and women of middle age, sufficiently old to have some experience and sufficiently young to be put in touch with present-day knowledge. We have to find out how to get the best out of every stage of human life at the same time as the span of human life itself is being extended.

The importance of education in the middle ages of man would do something to mitigate the present practice of giving the greatest importance to the elderly, largely by the automatic working of the principle of seniority. This results at present in the running of nearly all countries by people entirely incompetent to understand the new scientific basis on which the economies are based, and the forcing of people who do know enough science to spend too much of their time trying, usually vainly, to persuade virtual ignoramuses that they must do something in their own interests.

PLANNING

Opposition to the idea of planning science is a hangover from that of free-enterprise industry. It is now long out of date for, in fact, science is already planned in certain fields, but planned in the most unsatisfactory and self-defeating way. Once science became a large-scale enterprise the dominant system of science planning was that of the scientific project and the grant to provide money for it. A team research, to be effective, must be sufficiently large and costly. The normal way of financing it, is not to order it to be done, but to leave it open for bids or what might be called scientific tenders. Then, those scientists who may be competent, but certainly have to be businesslike and trained in sales methods and lobbying, put forward schemes for research and secure a budget for a certain number of years. The *results* of science, except in the military field, are more or less public and are subject to public criticism. The *planning* of science, however, is private and only subject to the criticism of the particular committee, where prejudice and personal interests, not to say political considerations, have full sway.

It is about time that scientists themselves were allowed to take the matter in hand and have public discussions and propositions on the proper development of science. For instance, the question of the amount of money and the distribution of money for space science as against molecular biology has been the subject of a number of unofficial protests, but there is no body or no organization where the scientists can make their collective judgement and will felt in the matter.

At times, both in the USA and the USSR, the whole planning system of official civil scientific research has had to be recast, and still a proper solution has not been found. And yet, without such a solution the progress of science will become ever more expensive and ever more inefficient.

TOWARDS A SCIENCE OF SCIENCE

Two of the essential difficulties in the planning of science is that it is planned according to no ascertainable principle and that it is not subject to any operational research. There is still no science of science; yet by now we should have enough experience about planning and results at least to begin the study of such a science. The enormous effort that is going on all over the world in the development and utilization of science and the amount of information already available about what has happened in the first two-thirds of the twentieth century should furnish enough observations to build quite a respectable science on it. This means treating the history of science not, as at present, as memorials to great men or as an example of social influences, but considering it critically in relation to good and bad efforts to find out things or to use what has been found out. We ought to demand, not only how was this discovery made, but why was it not made before then and what would have been the course if history had gone differently.

For instance, if the casual conversation of Henri Poincaré and Bequerel in 1897 had not taken place, it might have been many years before radioactivity was discovered. Ultimately, it was bound to come, because there are many effects which are traceable to it, but it would have been much harder to interpret. If the discovery of radioactivity had been delayed the result of human history might have been quite different. The Second World War and atomic fission only came together in time by the merest accident. If the bomb had come four years earlier we should have had atom bombs fully in use all through the war instead of having them only when it was virtually finished. Alternatively, if atomic fission had been discovered a year or two later it could have been developed quite comfortably for several civilian uses and might never have given rise to the most horrible shadow of the Cold War.

A plea for the creation of the science of science means again the setting aside of a certain number of people to discuss and teach it, at least in universities. Essentially, this is the carrying out in the twentieth century of Bacon's plea in the seventeenth (pp. 442 f.). There is very great prejudice against the idea of a science of science on account of its connexion with religious and metaphysical controversy. But we should not allow this to interfere with something that is essential for the proper carrying out of science at all.

Perhaps the most important effect of the new scientific technical revolution is that it forces us to re-evaluate the basis of our morality, at any rate of our collective morality, with respect to science itself. In so far as it is realized that the progress of scientific research and its application have now become the prime causes of human improvement,

393. In modern industry and in science it is necessary now to bring new knowledge to those already in employment or with professional qualifications. A case presentation to senior staff and postgraduate students at the Royal Postgraduate Medical School, London.

all interference with that scientific research, both material and mental, becomes a crime. The stinting of money on scientific research and on scientific education is holding up the whole of progress and condemning, in some cases, tens of millions of people to quite unnecessary hardship and deprivation: lack of medical research condemns them to disease and death; lack of agricultural research condemns them to starvation. By working with less than the known optimum amount and quality of equipment, the levels of technical performance are lowered and production is less than it could be with the same effort. In simple language, economy on research is waste, and waste that the world at its present stage of development can no longer afford.

Mentally, the blockage of research by the maintenance of obsolete ideologies is even more pernicious. The whole history of science demonstrates that the greatest effort is required not so much in discovering new things as in breaking obsolete ideology sanctified by custom or religion. This is to be seen particularly on the boundaries between science and practice. In considering the aid to under-developed countries, for instance, we are allowed to talk about new techniques, but not to talk about the social and economic habits that force the continuance of old unscientific practices, such as the absence of land reform and persistence of plantation economies. At the same time so-called religious arguments are used to prevent state medicine in some countries and to protect the sacredness of cows in others. The most damaging of these obstructions, because it operates at the centres of science in the capitalist countries, is the general block which exists on any really serious social science, namely, on any science which recognizes the economic realities of a class-divided society. It may take some time longer for these implications of the scientific revolution to seep in. Only when they have, will it be possible to have any security for the advancement of mankind.

14.10 The World's Need of Science

The major conclusion that arises from a study of the place and growth of science in our society is that it has become too important to be left to scientists or politicians, and that the whole people must take a hand in it if it is to be a blessing and not a curse. This is no distant prospect. Thanks to the uses to which science has been put, first by an unregulated capitalism in the Industrial Revolution and now by monopoly capitalism, the whole situation of humanity on the globe has become extremely insecure. The world is threatened as it was never threatened before with the twin dangers of war and famine.

THE DANGER OF WAR AND HOW TO MEET IT

The discussion of the effort spent on war preparations and actual war and the particularly close way in which science and the scientists have been and are involved in it, has had to occupy, regrettably, too many of these pages (pp. 717, 831–48, 1257 f.). We have examined at least some of the multiple ways in which even the preparations for war distort and stunt the growth of science. Yet none of these is comparable to the effect of war itself with the weapons we already know to be available. A new war would not wipe out civilization but it would set it back by many years; hundreds of millions of lives would be lost and suffering and disease would be multiplied manifold (pp. 839 ff.).

The resources of the world, as they are used at the present time, are insufficient for its needs; to squander more of them in war might upset the balance catastrophically and lead to almost unlimited disaster. It is therefore a matter of absolutely first priority to put an end to this impending possibility of war. And it can be done given sufficient popular understanding and popular pressure. It is true that war and war preparations have become apparently essential conditions for the preservation of capitalist economy in its present phase. This is not so much because of the need to defend it against external and internal enemies, but as a means of securing the maximum profitable production of goods which cannot embarrass the market by needing to be used for other than destructive ends.

The maintenance of the state of tension which could justify an indefinite Cold War has proved increasingly difficult. The fact that more and more people are appreciating the suicidal character of the hydrogen bomb makes generals and even politicians hesitant to do more than go to the brink. Sooner or later the intrinsic absurdity of wasting so much effort on armaments and scientific war preparations in a world that dares not go to war will sink in and there will be a return to sanity.

Once the immediate threat of war is removed, the way is open to establish some agreed form of co-existence between the two great systems of government in the world – capitalist and socialist. This is already recognized as involving a substantial measure of disarmament, including a secure agreement prohibiting the use of all weapons of mass destruction: atom and hydrogen bombs and biological warfare.[8,9] It would further imply a resumption and great enlargement of trade between the two parts of the world and a full measure of cultural and scientific exchanges. The consequences of such a *détente* in capitalist countries might involve economic recession due to reduction of armament orders, but this could be temporary and be more than offset by increased investment in internal development and trade, especially with the undeveloped and socialist countries. There, stabilized peace and disarmament would allow for more effort to be devoted to consumer goods and development plans. More money would also be available for investment and assistance to undeveloped countries (p. 977).

In this change science would be the largest gainer. If any substantial fraction of the resources now made available for military research and development were made available for civil research the result would be an increase in means and manpower greater than has ever been known in the history of science. Not only that, but it would be possible to convert military into civil research establishments of the same general character as rapidly as the reverse change was achieved in the last war – in a matter of a few months at most.[8,11,285]

All these hopes turn on the ability of the peoples of the world to compel their governments to prevent the outbreak of a Third World War. Because the weapons which are being prepared have largely been forged by science, the scientist has a special responsibility and should be deeply concerned in all efforts to stop war and to remove its political and economic causes. This, in the present conditions of the world, is a complicated task. The scientists have first of all the responsibility of analysing the situation as best they can, especially where they have technical competence, and in the light of this of helping to inform their fellow citizens, and joining movements that seem to them to tend in the direction of a practical and durable peace.

THE DEFEAT OF FAMINE

Even without war, moreover, the prospects for humanity are precarious enough. We have already described how the enormous expansion of agriculture in the nineteenth century has had the result of producing an immensely larger amount of foodstuffs at the expense of a wastage of soil on a scale never before reached: a wastage which is now cutting very appreciably into the soil reserves of the entire planet. This has coincided with an increase in population which was set in motion by

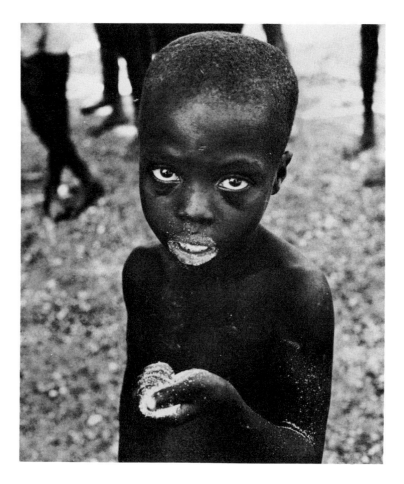

394. General and persistent famine is an ever present problem. Child eating raw flour distributed during a food shortage in Dahomey in 1962. Photograph taken for FAO.

395. The use of satellites to scan cloud cover over the world and to assist in the study of world weather conditions is a new and valuable technique. It may help in the future possibilities of weather control, and so in increasing rainfall in arid and unproductive areas. Photograph from Tiros IV.

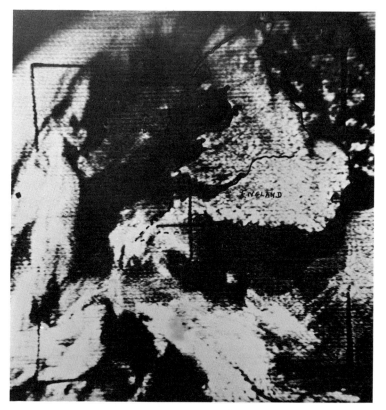

the need for labour in the first Industrial Revolution, and made possible by the improvements in agriculture, health, and transportation that accompanied it.

Starting in Britain and northern Europe, it has spread much further than the industrial centres into their fields of supply of raw materials. There the need for staple crop agriculture to produce food and fibres, now largely concentrated in tropical and sub-tropical regions, together with that for mining, creates a further demand for labour, increases the population and presses harder than ever on an almost static food supply. As we have seen (pp. 973 f.), this is not a simple, automatic consequence of biological factors, but rather one of an out-of-date and vicious system of plantation and semi-feudal agriculture in the service of big capital. But as long as it lasts famine – not the casual famine of one bad year located in one place, but general and persistent famine – will certainly be the result.

SCIENCE FOR WELFARE

The reversal of these trends is primarily an economic and a political problem. Only when the world is effectively secure against war, and is able to devote the enterprises of men and women to common welfare, is it worth considering in any detail the proper development and utilization of science. The job of using science for human betterment is also primarily political; that is, one that in the last resort must be settled by the people as a whole. But they cannot do it without the information that only the scientists possess. It is therefore the business of the scientist, at least for part of the time, to come out of his own

396. The use of artificial fertilizers is a vital matter if the world's food production is to be stepped up to keep pace with the increase in the world population. Testing fertilizer for caking properties in research laboratories at Billingham.

speciality and to work together with all other like-minded people in the different walks of life, that is with the professional, manual, and domestic workers, to ensure that we get a society where science can be properly used. But there is no reason why they should not do so, and it is here, where organization links scientists and non-scientists in common effort, that they can find their task.

Something of what this would mean in the actual production of good living conditions, in health, culture, and general happiness and effectiveness has already been told at the ends of Chapters 10, 11, and 13 dealing with twentieth-century science. To recapitulate: it would mean carrying through a new industrial and biotechnical revolution which would have the primary effect of getting rid not only of the need for all over-arduous and dangerous labour, but also of all monotonous factory or clerical labour, through the development of automatized production and control mechanisms.

POWER AND AUTOMATION AS THE SERVANTS OF MAN

All this would be on the basis of a full production and utilization of power which, once free from the stranglehold of bomb production, we now know is indefinitely available. It need not be used sparingly, but must be used wisely. Ample power means ample raw materials such as steel, other metals, plastics, fibres (p. 851): all that is needed for industries, transport, houses, or clothes. Raw materials mean factories which can make, through automation, all the goods we want. They also mean a rational transportation system for goods and one in which people can get where they want to go without getting in each other's way or spending their time driving about. There is no technical reason why traffic should not be already automatized. All this great accumulation of capital will take time, but apart from political obstruction it should not take much once the great reserves of productive and scientific capacity locked up in military preparations are released to do useful work. The most tragic irony of our times is that our greatest material and intellectual effort is devoted to keeping the world poor, ignorant, and frightened. Wealth, knowledge, and freedom are open to everyone in the world on a scale beyond the dreams of the richest of today. Only their narrow greeds, stupidities, and fears prevent them sharing to the limit of their desires in the common welfare.

THE TRANSFORMATION OF NATURE

At the same time the transformation of Nature, along lines indicated by the biological and geological sciences, will be undertaken with the

397. The wise use of natural resources can lead to a substantial improvement in conditions. The construction of hydro-electric generators at Cruachan, Loch Awe, Argyllshire.

use of heavy machinery, including possibly atomic energy. All the river basins of the world can be brought under control, providing ample power, abolishing floods, droughts, and destructive soil erosion, and widely extending the areas of cultivation and stock raising. The immediate effects of malnutrition and the fears of famine can be ended, and a large increase in population comfortably accommodated. Behind this lie possibilities of further extending the productive zone of the world to cover present deserts and mountain wastes and making full use of the resources of the seas, and beyond that again lie the possibilities of the microbiological and photochemical production of food.

THE POWER OF RESEARCH

These are no idle projects, but all realizable with the material, the people, and the knowledge we have already. But that is only the beginning of the promise of science. Its real value comes from revealing what we do not know, in the illimitable power of research which, fully manned, well supplied with a free and flexible organization, would be able to show both in extent and depth what humanity is capable of.

This is the 'light' of which Bacon dreamed 'which should in its very rising touch and illuminate all the border regions that confine upon the circle of our present knowledge' (p. 443). It would be most immediately useful where he would have most appreciated it, in banishing disease and lengthening life.

This economic, agricultural, scientific revolution will involve a reversal of the centralizing trend of industry under capitalism and its replacement by a balance of industry and agriculture all over the world. It will also involve an integration of science with the other productive forces. This implies a great increase in the scale of science itself. The effort expended in finding out scientifically what to do will tend to be as large as that used in doing it. The transformation should establish a new level of human control over its total environment. The agricultural and industrial revolutions of the past gave man technical and organizational control over his organic and inorganic environment. The present one should add control over the social environment and over the development of society itself. The achievement of a classless society is the first necessary condition to set the transformation going, but to complete it requires something altogether new: the acquiring by all of a capacity for conscious, integrated, social action.

SOCIAL RESPONSIBILITY

What such a transformation would mean for science and culture in general is intrinsically unpredictable. If we knew all the answers in advance there would be no use for science. We may reasonably infer, however, from the experience of history, that it would lead to an unprecedented activity in which new great works of human understanding and human construction would be created. Intellectual life, in which all the people would share, would become more closely knit, more linked with the practical tasks of humanity, and more responsible.

A new kind of society, conscious of its own development as a matter of the mutual responsibility of all its citizens, must needs evolve from this background its own morality, which, while incorporating all that has gone before, can reach new levels which we can now only glimpse at. The new technology, the new science, can no more be run with the old morality than it can be with the old economic and political systems. This will mean a level of individual and collective responsibility far higher than any reached in earlier times, limited as it was then to the traditional requirements of family and tribe.

An absence of knowledge meant automatically an absence of responsibility. When both the good and the ill fortune that befell men were

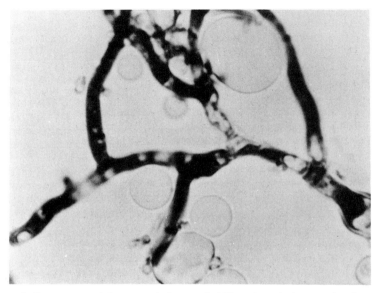

398. Recent research has proved the practicability of producing proteins from oil. A photomicrograph of moisture globules generated by micro-organisms cultured on gas oil at the Lavera refinery of the Société Française de Pétroles B.P. Such work as this may make a completely new source of protein available for the world at large.

not anything that they could understand or relate directly to their own action, it was not unreasonable to believe them to be controlled by other powers, by blind fate, or by helpful or malignant gods. In so far as individual responsibility was felt, it was indirect. If man could not control the elements he could at least attempt to control himself, and when a man failed to do so by defying the customs of the tribe or the commands of the gods, the consequences were visited in general misfortune. But the responsibility implied in the concepts of righteousness and sin is blind. Its acceptance can only lead to pious adherence to tribal rituals and taboos. Even in the age of capitalism, responsibility was individually limited and measurable simply in terms of money. If a man made his way honestly and supported his family, he had fulfilled his most essential responsibility to society. The miseries of boom and slump, of unemployment, slums, and war were no concern of his.

Time was when ignorance was pious. It suited the masters in a class-divided society well enough that the people should know no more than they needed for their work, and particularly that they should not inquire at all into the bases of society itself (pp. 1022 f.). With increasing know-ledge and experience, such blindness is now no longer acceptable, in-deed no modern industrial community can permit it and survive, and responsibility is again becoming collective and conscious. If they have done nothing more, events have taught us that men are no longer separate units: that their most apparently isolated acts are factors in a general social movement. The 'invincible' ignorance that was per-missible and indeed inevitable in past ages now becomes 'vincible' ignorance which can no longer be allowed. If some people refuse to see or to accept the implications of what they see, if – worse – through their control of education and the Press, they prevent others from seeing the significance of social action, they are effective enemies of society. Only when it is in nobody's interests to obscure social truths, that is only in a fully classless society, will it be possible for everyone to see un-obstructed the nature of the society he lives in.

Once a secure material basis has been won and natural resources are being wisely used, there will come the most important change of all – the liberation for the first time of the resources latent in human beings themselves. Over all the world, and no longer only for the limited classes and races, every child will be able to grow up with the full advantages of education. He or she will be able to make the fullest use of their abilities freely in the common service. In the process they will build and change the social forms through which they co-operate, but in full consciousness and by scientific discussion, no longer driven by arbitrary political masters or blind economic forces. In any society which is to be run by the people for the people, and in the struggle to secure such a society, the greatest knowledge of Nature and of society is an imperative requirement. To make it a reality implies the spread of a genuine popular education of a new kind, of which I have already spoken. In so far as that education becomes effective, so will the capacity of all people to use and take part in science; and the isolation of science from the rest of culture and from the people will come to an end.

It is through science, and only through science, that the transforma-tion of society to one free from exploitation can be brought about. Throughout the long reign of class-dominated societies, available tech-nique was never high enough to provide more than a small margin of production over subsistence, which was appropriated by the ruling class. Now, thanks to science, we can make that margin as big as we like, but misery and danger will remain the lot of man until science can be freely used, and not distorted for mean and destructive ends. In all previous class struggles one class simply took the place of another, and exploitation went on in a different form. In the transformation from capitalism through socialism to communism, that necessity will finally vanish, production will be ample enough to remove any need for proletarians or serfs. But there will still be a need for science, now not limited to a few specialists, but part of the life of the whole people.

There is plenty that still needs doing. The first and hardest step is to use our present knowledge to remove known evils. The second is to use research to find new means for removing evils that we cannot at present avoid, to cure diseases and maintain life and happiness for all. But beyond that there lie still further tasks, those of continuing and extending research to discover the unrecognized evils that we must in turn fight and destroy. Conversely, in a more positive way, we need to discover new good things, new materials, new processes and, most of all, new and effective bases of organization for social action. What this means, in effect, is that the task of human thought only begins in knowledge. Knowledge must result in constructive change before it can renew itself.

These anticipations may fitly close this book, since the purpose for which it was written was to search the past for clues to the future. In so far as my field was science, it was my business to stress the share of science in bringing about social changes and in creating social problems. I have tried also to show how science can help in solving them. They certainly cannot be solved without science, nor can they be evaded.

History has in the past been the record of human intentions, of human actions, and of events which were more often than not very different from the ends consciously aimed at. It was the field of action of forces that could only be dimly guessed at, and were far too easily identified with superior beings whose playthings were men. As we come to see more in history than this, as we begin to understand something of these forces and the laws they must obey, the events of history will become the results of conscious planning and achievement. With the discovery of the science of society, as Engels said, the true history of mankind begins.

Table 8

Science in History

In this table (next page) an attempt is made to draw together the results set out in earlier tables and to indicate in a way not given there the nature of their inter-action. The first three columns give the historical, social, and geographical back-ground. The time-scale, unlike earlier tables, is not uniform; periods of rapid advance such as the fourth century B.C. or the seventeenth century A.D. are given as much space as the four centuries between the fourth and eighth centuries A.D. This necessarily obscures the real unevenness of scientific and technical progress, but is inevitable from lack of space. The break in the year A.D. 1400 is also for convenience in printing. The third column naturally, especially in later time, gives only somewhat arbitrarily the main centres. It should be studied in conjunction with the maps. The main part of the table exhibits some of the interactions scientific and technical advances mixed up as they were in reality. Their nature is indicated by putting the more purely scientific aspects in italics. Some of the influences of technical and scientific factors on each other are indicated by arrows, though many are omitted to avoid too great a complication. Three main lines of continuous advance stand out. The first and longest, as far as science is concerned, is that of the Calendar, Astronomy, Navigation, Mechanics, splitting into various branches of physics. The second and even longer line includes the main sequence of mechanical technological advance, owing little to science up to the invention of the steam engine, but giving much to it. The third line is of advance in the arts of changing matter – Ceramics, Metallurgy, and Chemistry – in which the con-tributions of science and technology are far more mixed. Minor, more limited sequences, such as those of optics and electricity, are also indicated, but the relationships of the agricultural, medical, and biological sectors are too complex to be adequately indicated on this scale.

Interactions of Science and Technique

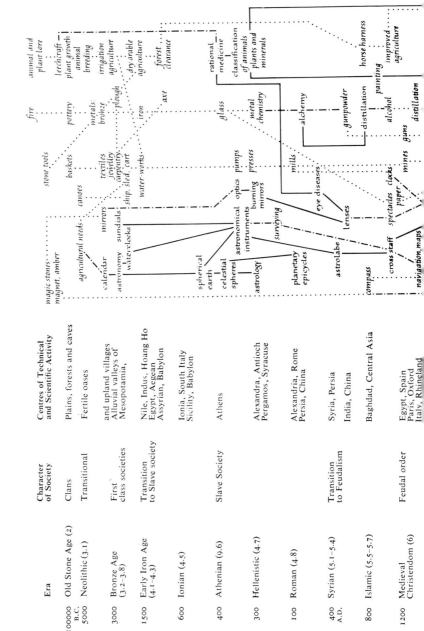

KEY

scientific connection ········· | interaction of science and technique – – – | technical connection ·········

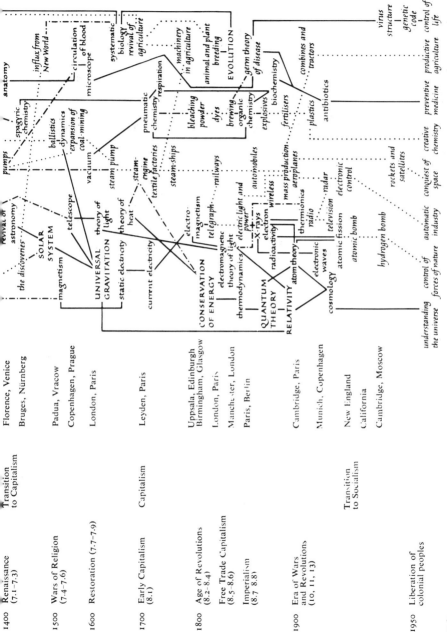

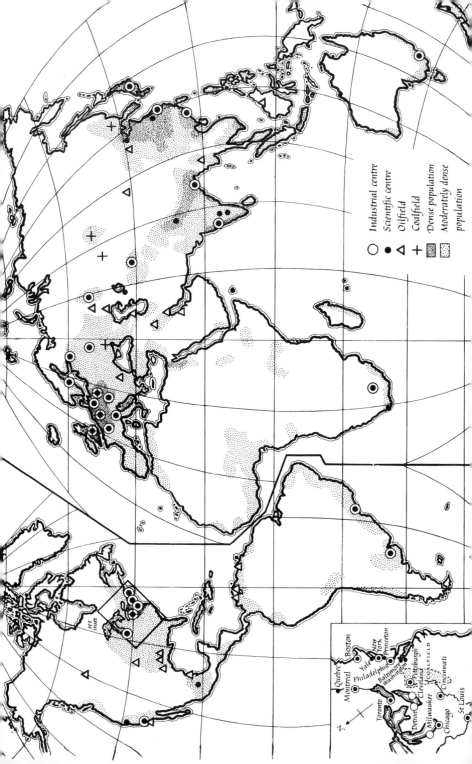

Industrial centre
Scientific centre
Oilfield
Coalfield
Dense population
Moderately dense population

○ ● ◁ + ▦ ▨

see inset

Quebec
Montreal
Toronto
Boston
Yale
New York
Princeton
Philadelphia
Baltimore
Washington
Pittsburgh
Cleveland
COALFIELD
Cincinnati
Detroit
Milwaukee
Chicago
St Louis

N

Map 5

The World Today

This map attempts to show the distribution of population, industry, and scientific effort throughout the world, corresponding to the discussion in Chapter 14, particularly on pp. 1249 ff. To bring out the main points, contrasts have been somewhat exaggerated. It will be seen that most of the population of the world occupies four main areas: Europe, North America, China with Japan, India with Indonesia. The industry of the world is, however, almost entirely concentrated in the first two areas – a situation not likely to last. The old population centres are developing industry rapidly and new centres of population and industry are growing up. That there is plenty of space for expansion is indicated by the large blank areas on the map. The smallness of the major industrial areas of north western Europe and north eastern America prevents the representation of their centres of production and research on the scale of this map. For the more detailed European distribution see Map 4. That for America is shown in the inset as well as in the key to leading universities and industrial cities. No key is provided for the map as a whole, as the location of most of the area is obvious.

Notes

For explanation see page 1013.

PAGE 1029. *Professor Gordon Childe published in 1956 an important and stimulating book in which he used his experience of anthropology and archaeology to examine values and showed them to be determined by a long and slowly evolving social tradition.[7,24] He found this particularly to apply to knowledge and 'truth', especially self-evident truth, which, far from being absolute, is always seen through the eye conditioned by society. He makes the valid point with regard to categories of thought that they, too, are traditionally determined through language. All languages are appropriate to a stone age technology when men controlled no motive power save their own muscles. There is little distinction between inanimate and animate objects, less between the latter and persons. Actions are confused with qualities and both with things. Childe defines Reality as a creative activity or process: 'The function of knowledge is practical – to guide action. The success of the human species, the only society known of knowers, . . . suffers to demonstrate that sufficient knowledge is attainable.' Perfect knowledge and absolute truth and the supersensuous world of ideas are as much fictions as any centaur. No one who has not at least tried to look as deep as Childe into the history of humanity has the right to question such conclusions.

PAGE 1047. *His task was to justify the seizure of a Portuguese treasure ship in the East Indies by the Dutch. They wanted to sell the goods in London at a time when England was not at war with Portugal. Grotius' argument was that, as the Dutch had expended so much money on ships, guns and powder, they were entitled to a reasonable return on their investment.

PAGE 1047. †In the Low Countries, the founders of the Dutch Republic spent years justifying themselves as really loyal subjects of the King of Spain. They objected only to the behaviour of his representatives in Holland, such as the Duke of Alba. Until the time of the Declaration of Independence the same attitude was maintained by the rebellious American colonies.

PAGE 1047. ‡Leibniz gave great thought to various devices for securing a permanent peace, foreshadowing the United Nations of our day.

PAGE 1057. *Such beliefs in the sanity of the common man and the decadence of nobility and clergy inspire much of pre-revolutionary thought from Goya's paintings to Mozart's operas. Beaumarchais' *Figaro* is the type of the new man on the point of emerging on the political stage. The equality of man was also a tenet of the faith of Freemasonry which permeated the society of all western European countries in the eighteenth century.

PAGE 1060. *This common judgement of Bentham seems to be only of his public face. Privately, as his recently published notes show, he entertained advanced socialist views.[7.10]

PAGE 1202. *In the case of the Congo, in the crisis that ensued, an effort was made to mitigate the chaos by referring the matter to the United Nations; but it is evident from what happened, particularly from the tragic deaths of Patrice Lumumba and the U.N. Secretary, Dag Hammarskjöld, that the United Nations' intervention was by no means impartial and that it effectively assisted the powers who held the principal wealth in the mines of Katanga.

PAGE 1225. *As Needham has shown, the pseudo-sciences in China proved most fruitful sources of genuine science, though they later outlived their usefulness.[8.47] Thus geomancy gave rise to the compass and in part geology. Various forms of drawing lots produced games like chess and cards, as well as leading to much mathematics, notably with the trigrams and hexagrams.

PAGE 1244. *The high point of Indian culture seems to be as early as Asoka in the third century B.C., though in art and science it was still progressing positively as late as the ninth century. In China the productive period is longer, though this may be simply because the records are better. The greatest intellectual ferment is in the fifth century B.C., and new ideas are emerging as late as the twelfth. Islamic flowering was brilliant but short; most is comprised between the ninth and the twelfth centuries. In all cases these civilizations seem to have passed their creative zenith and relapsed into traditional pedantry before their advance along the old lines was made impossible by the intervention of the latest comer, European culture, which began to make an impression only in the twelfth century.

PAGE 1253. *In a more recent study (1960) of what he calls the 'Research Revolution', L. S. Silk shows that the tempo of growth of scientific research in the United States, and consequently of the expenditure on science, had multiplied sixfold between 1946 and 1959.[7.146]

Bibliography to Volume 4

PART 7

1. ALLEN, G. C., *British Industries and their Organization*, 3rd ed., London, 1951
2. ALLEY, R., *Yo Banfa!*, Shanghai, 1952
3. ANGLO-AMERICAN COUNCIL ON PRODUCTIVITY REPORTS, London, 1949
4. AYER, A. J., *The Foundations of Empirical Knowledge*, London, 1947
5. AYER, A. J., et al., *The Revolution in Philosophy*, London, 1956
6. BAGEHOT, W., *The English Constitution*, London, 1867
7. BAUER, E., *L'Électromagnétisme Hier et Aujourd'hui*, Paris, 1949
8. BEARD, C. A. and M., *The Rise of American Civilization*, 2 vols., New York, 1927
9. BEER, M., *A History of British Socialism*, 2 vols., London, 1919
10. BENTHAM, J., *Economic Writings*, 3 vols., London, 1952, 1954
11. BERNAL, J. D., *The Freedom of Necessity*, London, 1949
12. BERNAL, J. D., *Marx and Science*, London, 1952
13. BERNAL, J. D., *World Without War*, 2nd ed., London, 1961
14. BEVERIDGE, Lord, Hansard: House of Lords, 20 May 1953, HMSO, London
15. BONNARD, A., 'Les Universités et la Paix', *Comprende*, no. 2, Venice, 1950
16. BRADY, R. A., *The Spirit and Structure of German Fascism*, London, 1937
17. BURN, D., *The Steel Industry 1939-1959*, Cambridge, 1961
18. BURNHAM, J., *The Coming Defeat of Communism*, London, 1950
19. BURNHAM, J., *The Managerial Revolution*, London, 1941
20. BURNS, C. D., *A Short History of Birkbeck College*, London, 1924
21. BUTTERFIELD, H., *The Origins of Modern Science*, 2nd ed., London, 1957
22. CHAMBERS, R. W., *Thomas More*, London, 1935
23. CHILDE, V. G., *Man Makes Himself*, London, 1939
24. CHILDE, V. G., *Society and Knowledge*, New York, 1956
25. CLARK, C., *The Conditions of Economic Progress*, 2nd ed., London, 1951
26. COATES, W. P. and A., *Soviets in Central Asia*, London, 1951
27. COLE, G. D. H., *Is This Socialism?*, London, 1954
28. COLE, G. D. H., *A Short History of the British Working Class Movement 1789-1927*, London, 1932
29. COLE, G. D. H., *Studies in Class Structures*, London, 1955
30. COLE, G. D. H., *World Socialism Re-stated*, London, 1956
31. COMTE, A., *The Positive Philosophy of Auguste Comte*, New York, 1858
32. CONANT, J. B., *The Education of American Teachers*, New York, 1963
33. CROSSMAN, R. H. S. (ed.), *New Fabian Essays*, London, 1952

34. DENT, H. C., *Education in Transition*, London, 1944

35. DE WITT, N., *Soviet Professional Manpower*, National Science Foundation, Washington, D.C., 1955

36. DOBB, M., *Political Economy and Capitalism*, 2nd ed., London, 1940

37. DOBB, M., *Some Aspects of Economic Development*, London, 1951

38. DOBB, M., *Studies in the Development of Capitalism*, London, 1946

39. DOVER, C., *Half-caste*, London, 1937

40. DRUMMOND, H., *Natural Law in the Spiritual World*, London, 1883

41. DUNHAM, B., *Giant in Chains*, Boston, 1953

42. DUNHAM, B., *Man Against Myth*, London, 1948

43. EATON, J., *Marx Against Keynes*, London, 1951

44. ENDICOTT, M. A., *Five Stars over China*, Toronto, 1953

45. ENGELS, F., *Anti-Dühring*, trans. E. Burns, London, 1934

46. ENGELS, F., *Dialectics of Nature*, trans. and ed. C. Dutt, London, 1940

47. ENGELS, F., *Ludwig Feuerbach*, London, 1941

48. ENGELS, F., *The Condition of the Working Class in England in 1844*, London, 1892

49. ENGELS, F., *The Origins of the Family, Private Property and the State*, London, 1946

50. ENGELS, F., *Socialism, Utopian and Scientific*, London, 1934

51. FABIAN SOCIETY, *Fabian Tracts*, nos. 1–188, 1 vol., London, 1884–1919

52. FARRINGTON, B., *Greek Science*, vol. 1, Penguin Books, 1944

53. FARRINGTON, B., *Greek Science*, vol. 2, Penguin Books, 1949

54. FARRINGTON, B., *Science and Politics in the Ancient World*, London, 1939

55. FARRINGTON, B., *Science in Antiquity*, London, 1936

56. FAST, H., *Citizen Tom Paine*, London, 1945

57. FISHER, H. A. L., *A History of Europe*, 3 vols., London, 1935

58. FISHER, Sir R. A., *The Design of Experiments*, 2nd ed., Edinburgh, 1937

59. FRAZER, Sir J. G., *Folk-Lore in the Old Testament*, 3 vols., London, 1919

60. FRAZER, Sir J. G., *The Golden Bough*, abridged ed., London, 1922

61. FREUD, S., *Totem and Taboo*, London, 1950

62. GALKIN, K., *The Training of Scientists in the Soviet Union*, Moscow, 1959

63. GUILLAN, G., *J.-M. Charcot, 1825–1893, His Life – His Work*, ed. and trans. P. Bailey, London, 1959

64. GALTON, Sir F., *Hereditary Genius*, London, 1869

65. HAMMOND, J. L. and B., *The Town Labourer, 1760–1832*, London, 1917

66. HAMMOND, J. L. and B., *The Village Labourer, 1760–1832*, London, 1912

67. HARTUNG, F. E., 'The Sociology of Positivism', *Science and Society*, vol. 8, 1944

68. HEARNSHAW, F. J. C., *A Survey of Socialism*, London, 1928

69. HILL, C., and DELL, E., *The Good Old Cause: The English Revolution of 1640–1660*, London, 1949

70. HMSO, *Higher Education: Report of a Committee under the Chairmanship of Lord Robbins 1961–63*, London, 1963

71. HMSO, *Scientific and Engineering Manpower in Great Britain*, London, 1956

72. HMSO, *Scientific and Technological Manpower in Great Britain 1962*, London, 1963

73. HMSO, *Technical Education*, London, 1956

74. HMSO, *Statistical Review of England and Wales*, London, 1952

75. HOBBES, T., *Leviathan*, ed. M. Oakeshott, Oxford, 1946
76. HOBSON, J. A., *Imperialism*, 3rd ed., London, 1938
77. HUGHES, E. R. (trans.), *The Great Learning and the Mean-in-Action*, London, 1942
78. HUTT, A., *This Final Crisis*, London, 1936
79. HUTTON, G., *We Too Can Prosper*, London, 1953
80. JAMES, W., *The Moral Equivalent of War*, New York, 1910
81. JENKINS, C., *Power at the Top: A Critical Survey of the Nationalised Industries*, London, 1959
82. JOHNSON, H., *China's New Creative Age*, London, 1953
83. KALDOR, N., and SILVERMAN, R., *A Statistical Analysis of Advertising Expenditure and the Revenue of the Press*, Cambridge, 1948
84. KEITH, Sir A., *Essays on Human Evolution*, London, 1946
85. KEYNES, J. M., *The General Theory of Employment, Interest and Money*, London, 1936
86. KING, B., *Russia Goes to School*, London, 1948
87. KOROL, A. G., *Soviet Education for Science and Technology*, London, 1957
88. LABOUR PARTY, *Equality*, London, 1956
89. LABOUR PARTY, *Labour's Colonial Policy: The Plural Society*, London, 1956
90. LABOUR PARTY, *Personal Freedom*, London, 1956
91. LABOUR RESEARCH DEPARTMENT, *Forty Years of the LRD*, London, 1952
92. LAWRENCE, F., 'Makarenko', *Modern Quarterly*, vol. 8, 1953
93. LENIN, V. I., *Essentials of Lenin*, vol. I, London, 1947
94. LENIN, V. I., *Imperialism*, London, 1948
95. LENIN, V. I., *Materialism and Empiro-Criticism*, London, 1948
96. LENIN, V. I., *The Revolution of 1905*, London, 1942
97. LENIN, V. I., *The State and Revolution*, London, 1947
98. LENIN, V. I., *The War and the Second International*, London, 1946
99. LILIENTHAL, D. E., *T.V.A.*, Penguin Books, 1944
100. LITTLE, I. M. D., *The Price of Coal*, Oxford, 1953
101. MACKENZIE, N., 'Poverty and Welfare', *New Statesman and Nation*, vol. 66, 1954
102. MAKARENKO, A. S., *The Road to Life*, 3 vols., Moscow, 1951
103. MALENKOV, G., *Report to the Nineteenth Party Congress*, Moscow, 1952
104. MANTON, S. M., *The Soviet Union Today*, London, 1952
105. MAO TSE-TUNG, *Selected Works*, vol. I, London, 1954
106. MARX, K., *Capital*, vol. I, London, 1946; vol. II, Chicago, 1885; vol. III, Chicago, 1909
107. MARX, K., *The Civil War in France*, London, 1942
108. MARX, K., 'A Criticism of the Hegelian Philosophy of Law', *Gesamtausgabe*, Berlin, 1927
109. MARX, K., *The Critique of the Gotha Programme*, London, 1946
110. MARX, K., and ENGELS, F., *Selected Correspondence*, London, 1943
111. MARX, K., *Selected Works*, vol. I, London, 1942
112. MARX, K., *Selected Works*, vol. II, London, 1942
113. MAYO, E., *The Social Problems of an Industrial Civilization*, Boston, 1945
114. MEHRING, F., *Karl Marx*, London, 1936
115. MILLS, C. W., *The Power Elite*, New York, 1956
116. MORGAN, L. H., *Ancient Society*, London, 1877

117. MORRIS, W., *The Letters of William Morris*, London, 1950
118. NEF, J. U., *War and Human Progress*, London, 1951
119. O'LEARY, DE L., *How Greek Science Passed to the Arabs*, London, 1948
120. ORR, Sir J., *Food, Health and Income*, 2nd ed., London, 1937
121. PACKARD, V., *The Hidden Persuaders*, Penguin Books, 1960
122. PALEY, W. S. (Chairman), *Resources for Freedom*, 5 vols., Washington, 1952
123. PARSONS, T., *The Social System*, London, 1952
124. PARSONS, T., *The Structure of Social Action*, Chicago, 1949
125. PASCALL, R., *Karl Marx: His Apprenticeship to Politics* (Labour Monthly Pamphlet), London, 1942
126. PASCALL, R., *Karl Marx: Political Foundations* (Labour Monthly Pamphlet), London, 1943
127. PAVLOV, I. P., *Lectures on Conditioned Reflexes*, 2 vols., London, 1941
128. PAYNE, G. L., *Britain's Scientific and Technological Manpower*, London, 1960
129. PEASE, E. R., *The History of the Fabian Society*, 3rd ed., London, 1963
130. PLATO, *Dialogues*, trans. B. Jowett, 3rd ed., 5 vols., Oxford, 1951
131. POPPER, K. R., *The Open Society and its Enemies*, London, 1945
132. PRICE, D. J. DE S., *Little Science, Big Science*, New York, 1963
133. RABELAIS, F., *Works*, Navarre Society, London, 1948
134. REITLINGER, G. R., *Final Solution*, London, 1953
135. ROBBINS, L., *An Essay on the Nature and Significance of Economic Science*, London, 1932
136. ROBERTSON, A., *The Bible and its Background*, 2 vols., London, 1949
137. ROBERTSON, A., *The Origins of Christianity*, London, 1953
138. ROBINSON, J., *The Accumulation of Capital*, London, 1956
139. ROLL, E., *A History of Economic Thought*, London, 1938
140. ROUSSEAU, J. J., *Oeuvres*, 22 vols., Paris, 1819-20
141. *Russia With Our Own Eyes* (British Workers Delegation), London, 1950
142. RYLE, G., et al., *The Revolution in Philosophy*, London, 1956
143. SAFONOV, V., *Land in Bloom*, Moscow, 1951
144. SARTRE, J.-P., 'At Vienna I Saw Peace', *Labour Monthly*, vol. 35, 1953
144a. SCIENCE AT THE CROSS ROADS, Papers presented to the International Congress of the History of Science and Technology, by delegates of the USSR, London, 1931
145. SHAW, G. B. (ed.), *Fabianism and the Empire*, London, 1900
146. SILK, L. S., *The Research Revolution*, New York, 1960
147. SIMON, B., *Intelligence Testing and the Comprehensive School*, London, 1953
148. *The Story of Ruskin College, 1889-1949*, Oxford, 1949
149. SMITH, A., *An Inquiry into the Nature and Causes of the Wealth of Nations*, 2nd ed., Edinburgh, 1846
150. SOCIALIST UNION, *Twentieth Century Socialism*, Penguin Books, 1956
151. SOONG, CHING-LING, *The Struggle for New China*, Peking, 1952
152. SPENCER, H., *First Principles*, London, 1862
153. SPENGLER, O., *The Decline of the West*, 2 vols., London, 1926, 1928
154. STALIN, J., *Collected Works (1907-13)*, vol. 2, London, 1953
155. STALIN, J., *Concerning Marxism in Linguistics* (Soviet News), London, 1950
156. STALIN, J., *Economic Problems of Socialism in the USSR*, Moscow, 1952
157. STAMP, L. D., *Land for Tomorrow*, Indiana, 1952
158. STOCKS, M. D., *The Workers' Educational Association*, London, 1953

159. STRACHEY, J. *Contemporary Capitalism*, London, 1956
160. STRAUSS, E., *Sir William Petty: Portrait of a Genius*, London, 1954
161. THOMPSON, E. P., *William Morris*, London, 1955
162. THOMSON, G., *Aeschylus and Athens*, London, 1946
163. THOMSON, G., *Studies in Ancient Greek Society*, London, 1949
164. TOYNBEE, A. J., *A Study of History*, 6 vols., Oxford, 1939
165. TREVELYAN, G. M., *Clio, A Muse*, London, 1949
166. TWAIN, M., and WARNER, C. D., *The Gilded Age*, London, 1885
167. TYLOR, E. B., *Anahuac*, London, 1861
168. VAILLANT, G. C., *The Aztecs of Mexico*, Penguin Books, 1950
169. VAIZEY, J. D., *The Economics of Education*, London, 1962
170. VAVILOV, S. I., *Isaac Newton*, Vienna, 1948
171. VEBLEN, T., *The Theory of the Leisure Class*, New York, 1899
172. VICO, G. B., *The New Science of Giambattista Vico*, trans. T. G. Bergin and M. H. Fisch, New York, 1948
172a. VOGT, W., *The Road to Survival*, London, 1949
173. WEBB, S. and B., *Soviet Communism: A New Civilization*, 2nd ed., London, 1937
174. WELLS, H. G., *The New Machiavelli*, London, 1911
175. WELTFISH, G., *The Origins of Art*, New York, 1953
176. WEST, A., *A Good Man Fallen Among Fabians*, London, 1950
177. WILLIAMS, F. E., *Soviet Russia fights Neurosis*, London, 1934
178. WILLEY, B., *The Eighteenth Century Background*, London, 1940
179. WILLEY, B., *The Seventeenth Century Background*, London, 1934
180. WILMOTT, C., *The Struggle for Europe*, London, 1952
181. WINSTANLEY, G., *Selections from His Works*, ed. L. Hamilton, London, 1944
182. WITTGENSTEIN, L., *Tractatus Logico-Philosophicus*, London, 1961
183. WOOLF, L. S., *After the Deluge*, 3 vols., London, 1931-53
184. YOUNG, J. Z., *Doubt and Certainty in Science*, Oxford, 1951
185. ZHDANOV, A. A., *On Literature, Music and Philosophy*, London, 1950
186. MYERS, C., *History of the Great American Fortunes*, 3 vols., Chicago, 1909-10

PART 8

1. ALLEN, J. S., *Atomic Imperialism*, New York, 1952
2. APPLETON, Sir E., 'Science for its Own Sake', *The Advancement of Science*, vol. 10, 1953
3. ASSOCIATION OF SCIENTIFIC WORKERS, *Science in Government and Industry*, London, 1962
4. ASSOCIATION OF SCIENTIFIC WORKERS, *Science and the Nation*, Penguin Books, 1947
5. AUGER, P., *Current Trends in Scientific Research*, Unesco, Paris, 1961
6. BAKER, J. R., *Science and the Planned State*, London, 1945
7. BAKER, J. R., *The Scientific Life*, London, 1943
8. BENOIT, E., and BOULDING, K. E., *Disarmament and the Economy*, New York, 1963

9. BERNAL, J. D., *Disarmament* (British Peace Committee), London, 1952
10. BERNAL, J. D., 'The Fourth Point and World Science', *Science and Mankind*, vol. 2, 1949
11. BERNAL, J. D., *The Freedom of Necessity*, London, 1949
12. BERNAL, J. D., *A Prospect of Peace*, London, 1960
13. BERNAL, J. D., 'Science and Human Welfare', *Science and Society*, vol. 20, 1956
14. BERNAL, J. D., 'Science in the Service of Society', *Marxist Quarterly*, vol. 1, 1954
14a. BERNAL, J. D., *World Without War*, 2nd ed., London, 1961
15. BICHOWSKY, F. R., *Industrial Research*, New York, 1942
16. BROWN, H. (ed.), *Science and the Creative Spirit: Essays on Humanistic Aspects of Science*, London, 1958
17. BUSH, V., *Science, the Endless Frontier*, Washington, 1945
18. CROWTHER, J. G., and WHIDDINGTON, R., *Science at War*, HMSO, London, 1947
19. CURTIS, C. P., *The Oppenheimer Case*, New York, 1955
20. DARWIN, C. G., *The Next Million Years*, London, 1952
21. DE WITT, N., *Soviet Professional Manpower*, National Science Foundation, Washington, D.C., 1955
22. DINGLE, H., *The Scientific Adventure*, London, 1952
23. DUBOS, R., *The Dreams of Reason: Science and Utopias*, New York, 1961
24. FEDERATION OF BRITISH INDUSTRIES, *Research and Development in British Industry*, London, 1952
25. FEDERATION OF BRITISH INDUSTRIES, *Scientific and Technical Research in British Industry*, London, 1947
26. HAILSHAM, Lord, *Science and Politics*, London, 1963
27. HMSO, *1954-55 Civil Estimates: Class IV*, London, 1954
28. HMSO, *DSIR Report for the Year 1951-2*, London, Cmd. 8773, 1953
29. HMSO, *Fifth Annual Report of the Advisory Council on Scientific Policy (1951-1952)*, Cmd. 8561, London, 1952
30. HMSO, *Higher Education: Report of a Committee under the Chairmanship of Lord Robbins 1961-63*, London, 1963
31. HMSO, *Present and Future Supply and Demand for Persons with Professional Qualifications in Physics*. Also in *Biology, Chemistry, Geology*, London, 1949
32. HMSO, *Returns from Universities, etc., in Receipt of Treasury Grant. Academic Year 1951-52*, Cmd. 8847, London, 1953
33. HMSO, *Royal Commission on the Civil Service* (1953). *Introductory Factual Memorandum on the Civil Service*, London, 1953
34. HMSO, *Science in the USA, 1952*, London, 1953
35. HMSO, *Science in the USA, 1953*, London, 1954
36. HMSO, *Scientific and Engineering Manpower in Great Britain*, London, 1956
37. HMSO, *Scientific Manpower*, Cmd. 6824, London, 1946
38. HMSO, *Sixth Annual Report of the Advisory Council on Scientific Policy (1952-1953)*, Cmd. 8874, London, 1953
39. HMSO, *Technical Education*, London, 1956
40. HMSO, *Working Party Reports: Cotton*, London, 1946
41. HMSO, *Working Party Reports: Wool*, London, 1947
42. HUXLEY, J., *Soviet Genetics and World Science*, London, 1949

43. MANCHESTER JOINT RESEARCH COUNCIL, *Industry and Science*, Manchester, 1954
44. MEES, C. E. K., and LEERMAKERS, J. A., *The Organization of Industrial Scientific Research*, 2nd ed., New York, 1950
45. NATIONAL MANPOWER COUNCIL, *A Policy of Scientific and Professional Manpower*, New York, 1953
46. NATIONAL SCIENCE FOUNDATION, *Federal Funds for Science*, Washington, 1953
47. NEEDHAM, J., 'L'Unité de la Science', *Archives Internationales d'Histoire des Sciences*, no. 7, 1949
48. NESMEYANOV, A. N., 'The Tasks of the USSR Academy of Sciences in Relation to the Fifth Five-Year Plan', *Bulletin of the Science Section: Society for Cultural Relations with the USSR*, October, 1953
49. PERLO, V., *American Imperialism*, New York, 1951
50. PFEIFFER, J. E., 'The Office of Naval Research', *Scientific American*, vol. 180, 1949
51. ROBINSON, J., *The Accumulation of Capital*, London, 1956
52. POLANYI, M., *The Logic of Liberty*, London, 1951
53. POLANYI, M., *Science, Faith and Society*, London, 1946
54. POWER, E. E., *The Wool Trade in English Medieval History*, London, 1941
55. PRICE, D. J. DE S., *Little Science, Big Science*, New York, 1963
56. PRICE, D. J. DE S., *Science Since Babylon*, London, 1962
57. RUSSELL, Earl, *Has Man a Future?*, London, 1961
58. SILK, L. S., *The Research Revolution*, New York, 1960
59. SNOW, Sir C., *The Two Cultures and the Scientific Revolution*, Cambridge, 1959
60. SPENGLER, O., *The Decline of the West*, 2 vols., London, 1926, 1928
61. STAMP, L. D., *Land for Tomorrow*, Indiana, 1952
62. STEELMAN, J. R., *Science and Public Policy*, vol. 4, Washington, 1947
63. TOYNBEE, A. J., *A Study of History*, 6 vols., Oxford, 1939
64. UNITED NATIONS, *Demographic Yearbook*, New York, 1950
65. UNITED NATIONS, *Economic Survey of Asia and the Far East*, New York, 1954
66. UNITED NATIONS, *Review of Economic Conditions in Africa*, New York, 1951
67. UNITED NATIONS, *Measures for the Economic Development of Underdeveloped Countries*, New York, 1951
68. UNITED NATIONS, *World Facts and Figures*, New York, 1953
69. VAVILOV, S. I., *Soviet Science: Thirty Years*, Moscow, 1948
70. WHO, *Annual Epidemiological and Vital Statistics*, Geneva, 1963
71. WORLD FEDERATION OF SCIENTIFIC WORKERS, *Science and Mankind*, no. 1, London, 1949

Note on the Illustrations

The choice of illustrations for Professor Bernal's *Science in History* has been based on the simple principle of providing additional illumination of the text. Since the author has taken so wide a canvas on which to display his analysis, the range of illustrations has accordingly been made as broad as possible. However, science has not always been illustrated at every stage in its history and from some periods, of the few illustrations which may have existed, little or no evidence has survived to the present day, in consequence certain problems had to be solved if gaps were to be avoided. For example, virtually no original material remains of Greek science, and the scientific texts that we have are copies or translations made in later centuries. In such cases, later sources have been used if, as often happens, they make the point; Greek ideas continued for so long in western Europe that it is often still valid to use material from printed books.

In this book, where both science and the interplay of social conditions are discussed, the pictures could not always be chosen as direct illustrations of the text. In every case it is hoped that the full captions will enable the reader to see why a picture has been chosen and appreciate its relevance, whether as allusion or analogy, by comparison or even as a comment. No attempt has been made to illustrate Professor Bernal's introductions to the various sections of his book, since this would have caused too great a mixture of subjects and historical periods. By confining illustration to the main body of the text, some degree of chronological order has been possible.

The choice of each picture has depended on a number of factors: its relevance to the text, the quality of the illustration itself, its power to provide additional visual or factual information and, of course, its aesthetic appeal. Here and there diagrams have been used, but in every case they are of historical significance. In volume 1, except for the need to cast the net wide for material about Greek science, the illustrations are comparatively straightforward. Volume 2 has almost illustrated itself. Volume 3, dealing primarily with modern scientific resesrch, is again straightforward, but volume 4 has presented some problems, in that its theme – the social sciences in history – is so wide, and that some of the concepts cannot be illustrated directly. The solution adopted has been to try, in one way or another, to complement the spirit of the text. Sources of illustration have been given wherever possible, in a separate acknowledgements section on p. 1329.

My thanks are due to Mr Francis Aprahamian for his helpful advice, and especially to my wife, whose assistance and extensive library of illustrations has proved invaluable.

COLIN A. RONAN
Cowlinge, Suffolk
June 1968

Acknowledgements for Illustrations

For permission to use illustrations in this volume, acknowledgement is made to the following: Aerofilms Ltd, 380; Aldus Books Ltd, 371a; the British Broadcasting Corporation, 381, 382, 391, 392; British Hydrocarbon Chemicals Ltd, 375; the Trustees of the British Museum, 314, 333; British Petroleum Co. Ltd, 364, 374, 398; China Features, 354; R. J. Chinwalle, 307; the Daily Mail, 344; Edwards High Vacuum Ltd, 369b; Fox Photos Ltd and the Communist Party, 339; I.F.A., 348; the Government of India, 376; Imperial Chemical Industries Ltd, 368b, 396; Ku Sung-nien, 362; the Mansell Collection, 308, 311, 317, 327; the Marx Memorial Library, 334; the Ministry of Health, 387 (crown copyright); the National Film Archive, London, 332, 342, 361; the Trustees of the National Gallery, 310; the National Physical Laboratory, Teddington, Middlesex, 379; Novosti Press Agency, 360; Radio Times Hulton Picture Library, 340, 341; the Ronan Picture Library, 309, 312, 313, 315, 316, 318, 320, 321, 322, 324, 325a, 325b, 326, 328, 329, 330, 331, 335, 336, 337, 343, 363, 368a, 369a, 370, 384, 386, 397; Professor J. Rotblat, 350; Ruskin College, Oxford, 346; Science Information Service, 345, 372, 377, 378, 394, 395; Shell Petroleum Ltd, 373; Shelter, 365 (photograph by Penny Tweedie); the Society for Cultural Relations with the U.S.S.R., 323, 338, 351, 352, 353, 355a, 355b, 357, 358, 359, 366, 367a, 367b, 383; United Kingdom Atomic Energy Authority, 371b; United Kingdom Atomic Energy Authority and Science Information Service, 389; 'Visual Education' and the National Committee for Audio-Visual Aids in Education, 389; Workers' Educational Association, 347; S. B. Williams, 393; the World Federation of Scientific Workers, 390. Photograph 345 is copyright the British Thomson Houston Co. Ltd.

The publishers would like to thank Mrs Sheila Waters for drawing Map 5.

Name Index

Bold figures indicate main reference

Subject Index

Bold figures indicate main reference